Popular Lectures in Mathematics

Survey of Recent East European Mathematical Literature

A project conducted by
Izaak Wirszup,
Department of Mathematics,
the University of Chicago,
under a grant from the
National Science Foundation

Yu. A. Shreider

What Is Distance?

**Translated and
adapted from the
Russian edition by
Leslie Cohn and
Harvey Edelberg**

**The
University of Chicago
Press**
Chicago and
London

The University of Chicago Press, Chicago 60637
The University of Chicago Press, Ltd., London

International Standard Book Number: 0–226–75498–7
Library of Congress Catalog Card Number: 74–5729

Contents

Introduction

This book is an elaboration of a course given by the author at Moscow University for pupils in the ninth and tenth grades. In it we discuss the development through abstraction of the general definition of distance and introduce a class of spaces in which the notion of distance is defined, the so-called metric spaces. It will be evident from our discussion that the general concept of distance is related to a large number of mathematical phenomena.

With the aid of the concept of distance, it is possible to study problems concerning the "shortest" path between two points on a surface, the geometric properties of multidimensional spaces, methods of "noise" reduction in the coding of information, and methods of "smoothing" errors in the results of empirical measurements, as well as many other such topics.

The concept of "distance," moreover, is a good illustration of the role played in mathematics by the generalization of specific ideas, the results of which at times find some rather unexpected applications. Other good examples of such generalizations which have been found indispensable to many areas of mathematics may also be cited: the notions of *function*, *limit*, *space*, and *transformation*, as well as the less familiar concepts of *isomorphism*, *group*, *ring*, and so on. Of these examples, however, the concept of *distance* seems most suited to the type of elementary discussion required by the inexperience of our audience, a consideration which is the chief motivation for our choice of this particular topic. Our aim is to demonstrate by means accessible to a wide range of readers the way in which one fruitful idea can shed light on a wide variety of mathematical questions and, at the same time, serve as a source of new results and insight in some particular field of knowledge. This situation, characteristic of all of the sciences, appears quite often in mathematics in particularly striking ways, making possible a clear understanding without the necessity of mastering a myriad of confusing details. The material for this book has been chosen with this general idea in mind.

The first four chapters are intended to expose the reader to the generalization of the ordinary geometric definition of distance and to the illustration of the generalized concept via concrete situations. Chapter 5 describes the so-called space of information, a concept that plays a major role in the theory of information and the general theory of communication. Chapter 6 deals with methods of coding information which allow that information to be relatively unaffected by errors in the process of transmission. Since in all real communications devices, errors occur in a number of ways, such methods of coding are essential for modern systems of communication and control. For example, in the transmission of photographs from the far side of the moon by a Soviet space vehicle, error-reducing methods of codification had to be used. It is important to note that each of these methods involves the use of the generalized concept of distance in the space of information.

The material in chapter 7 is somewhat more complicated; there we deal with an important class of spaces to which the notion of distance is common. Chapter 8 describes the application of the generalized concept of distance to the problem of "smoothing" errors in the results of empirical measurements—the problem of finding a mathematical process which will nearly eliminate the effect of error in experimental data. This chapter is essentially an exposition of the method of *least squares*. Some knowledge of differential calculus is necessary for an understanding of this chapter. The reader who has not had the necessary background may omit this section.

In the final chapter, the possibility of further generalization of the concept of distance is examined. In this chapter I wish primarily to show that it is not necessarily true that all generalizations possess interesting properties. It is not always easy to develop a good generalization of a mathematical concept. At the core of any worthwhile generalization are some essential properties of the real world. In particular, the concept of distance is important because many essential properties of real objects are related to their mutual disposition, which can frequently be characterized by a properly defined concept of distance. For example, although it is impossible to describe the electrons of an atom as point masses, quantum mechanics is nevertheless able to determine the "distance" between the two energy states of electrons. This "distance" is related conceptually to the "distance" defined in the so-called l_2 space discussed in chapter 7.

I shall consider my task complete if this book is able to give the reader a satisfactory understanding of the ideas mentioned above.

I wish to take this opportunity to express my gratitude to I. M. Yaglom, who has provided much valuable advice concerning the improvement of this manuscript.

1 The Definition of Mathematical Concepts

At first glance, the title of this book may seem surprising. Every schoolboy, it would seem, knows what distance is. Even a person who has completely forgotten his high-school geometry and who cannot accurately formulate a definition of distance would be quick to assert that he knows very well what distance is.

But, in fact, the matter is much more complicated.

The word *distance* can take on different meanings depending upon what particular space one is talking about. We are about to see that this is true even in situations with which we are well acquainted.

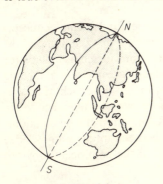

Fig. 1.1

In the Euclidean plane and in ordinary three-dimensional Euclidean space, the distance between two points M and N is defined as the length of the line segment MN joining those points. When dealing with distances between geographical loci on the surface of the earth, however, we usually have in mind the length of the smaller arc of the great circle joining those localities. The difference between these two meanings of distance becomes particularly noticeable if we calculate the distance between the north pole N and the south pole S (see fig. 1.1). The ordinary (Euclidean) distance between the poles is equal to the diameter of the earth, approximately 8,000 miles. The distance between the poles along the surface of the earth is, however, greater than this by a factor of $\pi/2$; it is about 12,500 miles.

To this example one might add that, in commerce, even the means of the transportation to be used must be taken into account in the estimation

1

of distances between cities. For example, the distance between two points by car may differ from the distance by train.[1]

We can obtain another example of distance if we consider points in rugged terrain and define the distance between two such points as the time necessary for someone on foot to travel from one point to another.

It is clear that this distance has nothing in common with the length of the line segment joining two points, for the straight line, in general, is not the best or most possible path. Indeed, a foot traveler will calculate the distance between two points by the time he spends in travel between them.

Despite differences among these means of measurement, however, it is evident that all meanings taken on by the word *distance* have something in common. A measure of "how far apart" two objects are is always indicated. Thus, one may suppose that there exists some common definition of distance which has various interpretations in various concrete situations. Such a general definition will be formulated in chapter 3. But first we shall consider what, in general, is necessary for the definition of a mathematical concept.

Modern mathematics is the language of natural science. Underlying the most important mathematical ideas are spatial-temporal facts about the world in which we live. However, the relationship between these facts and the corresponding mathematical ideas is sometimes very complicated.

In every branch of mathematics are some fundamental concepts which are related in our minds to certain physical images. Some of the fundamental properties of these concepts are formulated as axioms (or postulates); "truths" that are not proved but accepted as a starting point. All of the remaining propositions of the given branch of mathematics are derived logically from these axioms without reference to the properties of the physical world. The very formulation of a set of axioms expresses to some degree the relationship between intuitive knowledge of properties associated with these ideas and the empirically obvious properties of their physical forms.

Some of the most important concepts involved in geometry are the ideas of *point, straight line, plane, space,* and so on. In a systematic geometry course it is necessary to develop a list of the most basic properties of these concepts in the form of a set of axioms, the basis on which the whole structure of geometry is built.[1]

Some of the principal concepts involved in algebra are those of sets of *numbers* and operations on these numbers. For example, the structure

---

1. The first to fashion such an exposition of geometry was the ancient Greek mathematician Euclid (fourth–third century B.C.).

of the *integers, rational numbers, algebraic numbers, real numbers, complex numbers,* and so on, are studied.

In each of the five number systems specifically mentioned above, one can verify that certain fundamental laws concerning operations on numbers are satisfied. These are the commutative law for addition ($a + b = b + a$), the associative law for addition ($[a + b] + c = a + [b + c]$), the commutative law for multiplication ($ab = ba$), the associative law for multiplication ($[ab]c = a[bc]$), the distributive law ($[a + b]c = ac + bc$), and the rules $a - a = 0, a \times 1/a = 1$ for $a \neq 0$, which characterize the relationship between the principal operations (addition and multiplication) and their inverses (subtraction and division). All of these laws are satisfied in the number systems listed above to which they apply. However, it is not always the case that a given operation is defined in a given number system. Division is not always possible within the integers and, therefore, is not well-defined as an operation on the set of integers. If a number system contains only positive numbers, subtraction is not always possible. As it happens, certain rules for algebraic transformation of various expressions depend only on the properties listed above. For example, all of the rules for the solution of first-degree equations and systems of such equations are based upon these laws and upon the possibility of carrying out the operation of division.

It turns out, in fact, that it is possible to study many properties of various number systems as consequences of the general theory of systems on which defined operations (called addition and multiplication) satisfy the properties listed above. Such systems are termed *commutative rings* or *fields* in modern algebra (depending on whether it is always possible to carry out division).[2]

It is possible to view the rules for transformation of expressions and for solution of equations in the case of an arbitrary field or ring and to look at the rules normally developed in high-school algebra as special cases.

In contemporary algebra, rings and fields are usually studied as generalizations of number systems studied in high school. The basic properties of operations that can be carried out for integers or for rational numbers are set down as a starting point, and facts that may be derived logically using these properties alone are studied.

In taking this approach, mathematicians are interested not only in discovering new properties of the physical world and establishing relationships among these properties, but also in clarifying properties of

2. For a definition of *ring* and *field* see Birkhoff and MacLane, *A Survey of Modern Algebra* (New York: Macmillan, 1965).

"imaginary" worlds developed by using axioms similar to those of the number systems most closely related to physical reality.

This facet of mathematics is no less important than the possibility of describing the physical world. The Russian mathematician N. I. Lobachevskii, by altering one of Euclid's postulates, created an "imaginary" geometry, which, long afterwards, served as the basis of new physical concepts of the universe arising from Einstein's development of the theory of relativity.

In this book we shall study one of the most important of mathematical concepts—the concept of *distance*.

Our first attempt will be the listing of those properties of distance which are essential to elementary geometry. With these laws as our basis, we shall derive the definition of a so-called *metric space* and study various examples of such spaces. We shall see that such a specifically mathematical approach to the study of certain concepts from the point of view of a generalized concept reveals many interesting facts.

This approach—the creation of generalized concepts and the attempt to describe physical realities with the aid of these concepts—is characteristic of modern mathematics and its fields of application.[3] From this point of view, the concept of distance provides a good example of the fruitfulness of such an approach.

3. We must not overlook the role played in cybernetics by such generalized mathematical concepts as *information, automata theory,* and *algorithm.*

2 Distance and Its Properties in Elementary Geometry

We hope to arrive at a general definition of distance by generalizing the properties of "ordinary" distance in three-dimensional Euclidean space. Therefore, we shall first attempt to list the fundamental properties of ordinary distance.

Let us agree to denote the distance between two points M and N in three-dimensional space—the length of the line segment MN—as $d(M, N)$.

This notation emphasizes the fact that the distance between M and N is a real number which is completely determined by points M and N. In other words, distance is a real-valued function of pairs of points. If we characterize each point by an ordered triple of coordinates, say $M = (x, y, z)$ and $N = (x_1, y_1, z_1)$, then distance in three-space becomes a function of six variables:

$$d(M, N) = F(x, y, z, x_1, y_1, z_1).$$

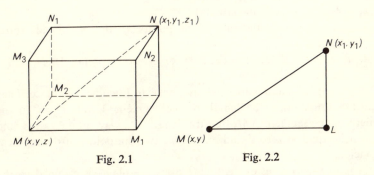

Fig. 2.1   Fig. 2.2

With the aid of figure 2.1, one can derive a closed algebraic expression for this function. Pictured is a parallelepiped with sides parallel to the coordinate axes. We know that the square of the length of the diagonal of a parallelepiped is equal to the sum of the squares of the lengths of its sides. Consequently,

$$MN^2 = MM_1{}^2 + MM_2{}^2 + MM_3{}^2$$
$$= (x - x_1)^2 + (y - y_1)^2 + (z - z_1)^2 \, ,$$

or

$$d(M, N) = \sqrt{(x - x_1)^2 + (y - y_1)^2 + (z - z_1)^2} \, . \qquad (2.1)$$

It is even simpler to calculate the distance between the points $M = (x, y)$ and $N = (x_1, y_1)$ in the Euclidean plane (see fig. 2.2). For this calculation, we need only note that the length of the line segment ML is just $|x - x_1|$, and, similarly, that the length of the line segment LN is $|y - y_1|$. By the Pythagorean theorem,

$$MN^2 = ML^2 + LN^2 \, ,$$

so that

$$d(M, N) = \sqrt{(x - x_1)^2 + (y - y_1)^2} \, . \qquad (2.2)$$

Despite the importance of equations (2.1) and (2.2), the properties of distance that we shall need can be obtained without the use of a coordinate system.

These properties can be formulated as follows:
1. $d(M, N) = d(N, M)$ (symmetry).
2. $d(M, N) \geq 0$ (nonnegativity).
3. $d(M, N) = 0$ if and only if the points M and N coincide (non-degeneracy).
4. $d(M, N) \leq d(M, L) + d(L, N)$ for arbitrary points M, N, and L (the *triangle inequality*).

Properties 1, 2, and 3 are obviously basic to Euclidean distance. They indicate simply that the length of the segment MN is equal to the length of the segment NM, that this length is always nonnegative, and that it is equal to zero if and only if the two endpoints of the segment coincide.

Property 4 becomes evident if we draw the plane determined by the points M, L, and N (and, therefore, containing the triangle MLN) (fig. 2.3). Property 4 then indicates only that the length of side MN does not exceed the sum of the lengths of the remaining sides of the triangle (hence the name *triangle inequality*). In other words, the straight line segment *MN* is the shortest path joining the points *M* and *N*.

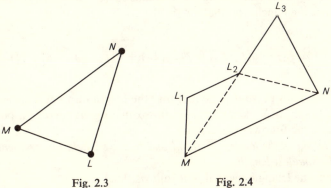

Fig. 2.3   Fig. 2.4

In fact, the triangle inequality becomes a *strict inequality* [$d(M, N) < d(M, L) + d(L, N)$)] in Euclidean three-space when we introduce the added restriction that *L* does not lie on segment *MN*. Hence, we can conclude that the length of segment *MN* is strictly less than the length of a broken line consisting of an arbitrary finite number of segments whose union joins the points *M* and *N*. In order to justify this conclusion (fig. 5), we shall repeatedly decrease by one the number of segments in the broken line, until, finally, only two segments remain. At each step in this process the length of the broken line will be strictly lessened until we reach the segment *MN* itself. Thus, in figure 2.4 we go from the broken line $ML_1L_2L_3N$ to the broken line ML_1L_2N, then to the broken line ML_2N, and finally to the segment *MN*. Each time the length of the broken segment decreases, and thus the length of the original broken line is strictly greater than the length of the segment *MN*.

Table 2.1

Broken line	Its length	Application of the strict triangle inequality
$ML_1L_2L_3N$	$d(M, L_1) + d(L_1, L_2) + d(L_2, L_3)$ $+ d(L_3, N)$	$d(L_2, L_3) + d(L_3, N) > d(L_2, N)$
ML_1L_2N	$d(M, L_1) + d(L_1, L_2) + d(L_2, N)$	$d(M, L_1) + d(L_1, L_2) > d(M, L_2)$
ML_2N	$d(M, L_2) + d(L_2, N)$	$d(M, L_2) + d(L_2, N) > d(M, N)$
MN	$d(M, N)$	

Let us note that in this deduction we use only the strict triangle inequality for Euclidean space. This can be best illustrated by table 2.1. From this table it is evident how, by replacing the sums in the second column by lesser sums using the inequalities from the third column, we arrive at the conclusion that

$$d(M, L_1) + d(L_1, L_2) + d(L_2, L_3) + d(L_3, N) > d(M, N) .$$

If, in addition, we use the fact that the length of a curve is the limit of the lengths of broken segments approximating the curve, it is possible to prove the following assertion:

Of all the paths joining points M and N, the straight line segment MN has the smallest length.

From the triangle inequality it follows that

$$d(L, N) \geq d(M, N) - d(M, L) . \tag{2.3}$$

Let us emphasize that equality holds in the triangle inequality for our three-dimensional example if and only if the points M, N, and L lie on the same straight line and L is located "between" M and N (that is, L lies on the segment MN).

Let us now examine a distance function on the surface of a sphere S of radius r.

We define the distance between two points M and N on the surface of a sphere as the length of the smaller arc of the *great circle* passing through the points M and N. Let us recall that a circle lying on the surface of a sphere is called a great circle if its center coincides with the center of the sphere. In other words, a great circle lies on the plane passing through the points M, N, and O (O being the center of the sphere). It follows that each pair of distinct points M and N uniquely determines a great circle, since three distinct points uniquely determine a plane. The distance $d_S(M, N)$ defined in this way clearly satisfies properties 1, 2, and 3. It is not difficult to see further that for arbitrary points M and N on the sphere,

$$d_S(M, N) \leq \pi r , \tag{2.4}$$

with equality holding only for points M and N lying at the endpoints of a diameter of the sphere (for example, the North and South Poles).

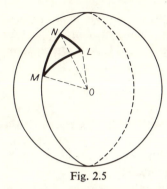

Fig. 2.5

To verify the fourth distance property, it is necessary to examine the spherical triangle *MLN* (fig. 2.5). (The point *O* is the center of the sphere.)

It is clear that

$$d_S(M, N) = r\alpha, \qquad d_S(L, N) = r\beta,$$
$$d_S(L, M) = r\gamma,$$

where α, β, and γ are the radian measures of angles *MON*, *LON*, and *LOM*, respectively.

It is well known that in such a trihedral angle none of the planar angles exceeds the sum of the two other planar angles; in particular,

$$\alpha \leq \beta + \gamma. \tag{2.5}$$

Multiplying both sides of this inequality by the radius *r*, we obtain

$$r\alpha \leq r\beta + r\gamma,$$

or

$$d_S(M, N) \leq d_S(M, L) + d_S(L, N), \tag{2.6}$$

the inequality we set out to establish.

Thus, all of the fundamental properties of ordinary distance are satisfied by the spherical distance $d_S(M, N)$.

It is easy to show that equality holds in inequality (2.6) if and only if two conditions are satisfied: first, that the point *L* is located on the same great circle as the points *M* and *N*; and second, that *L* lies "between" *M* and *N*—on the smaller arc of the great circle determined by *M* and *N*.

This follows from the fact that inequality (2.6) becomes an equality only when equality holds in (2.5). But this can occur only when the trihedral angle degenerates into a planar one—that is, when the points *M*, *N*, and *L* lie on a plane passing through the center *O* of the sphere and ray *OL* is located between rays *OM* and *ON*. But this implies that the point *L* lies on the smaller arc of the great circle joining points *M* and *N*.

It is evident that the smaller arc of the great circle joining points *M* and *N* possesses properties analogous to those of the straight line segment in ordinary (nonspherical Euclidean) geometry. In particular, (1) through a pair of arbitrary distinct points there passes exactly one such arc (with the exception of the case where the points *M* and *N* lie at the endpoints of a diameter of the sphere—that is, where they are *antipodal*—in which case both arcs of any great circle joining M and N are of equal length, and there are infinitely many such circles); (2) for any point L lying on such an arc joining the points M and N, the equation

$$d_S(M, L) + d_S(L, N) = d_S(M, N)$$

holds.

Let us note at this point an important extension of a fact proven in Euclidean three-space. For ordinary distance we have shown that the length of any broken line joining two points M and N is greater than the distance between the points M and N, that is, than the length of the segment MN. Here we base our reasoning only on the triangle inequality and on the fact that equality holds only if the points M, L, and N lie on the same segment (with L "between" M and N). Since the triangle inequality is also true for the distance function we have defined on the sphere, with the ordinary line segment corresponding here to the smaller arc of a great circle, it is apparent that an analogous assertion is true on the sphere: If the points M and N are joined by a broken sequence of arcs of great circles (fig. 2.6) in which successive arcs are joined by a common endpoint, then the total length of such a "spherical broken line" is greater than the distance $d_S(M, N)$.[1]

We suggest that the reader write up a full proof of this assertion in analogy with the proof for ordinary distance carried out above. This assertion can easily be generalized (using limit arguments) in the following form: *The length of the smaller arc of the great circle joining the points M and N is less than the length of any other path on the sphere connecting these points.*

Thus, we have examined two examples of distance and determined that their fundamental properties are the same. Auxiliary properties

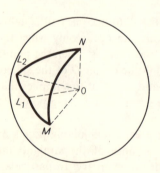

such as (2.4) (p. 8), properties peculiar to the particular example, play a much smaller role.

Therefore, our next step will be to take the fundamental properties of distance (1, 2, 3, and 4) as axioms and to study various *spaces* in which a *distance* satisfying these axioms is defined. In this chapter we have examined two elementary examples of such spaces: ordinary Euclidean three-space and the surface of the sphere.

Fig. 2.6

1. We must assume here, of course, that this sequence actually is "broken"; that is, that it does not lie entirely on the smaller arc of the great circle joining the points M and N.

3

The Definition
of a
Metric Space
and of Distance

We shall begin with an explanation of what a *set* is. Like the notion of *point* in geometry, the concept of *set* is fundamental and yet difficult to define. The word *set* is used in mathematics to indicate a collection of objects called *elements* of the set.

The concept of set has important applications in any situation where a general property is assigned to certain objects. When these objects fall into some class according to some sort of rule, they form a set. We shall say that a set *contains* each of its elements, and that each element of a given set is *contained* in it. A set is considered specified if for any arbitrary object it is possible to determine whether or not it is contained in the given set.[1]

Let us consider, for example, the set of all integers. The sun is not contained in this set as it is not a number but an object of an entirely different sort. The number π is not contained in this set, for it is not integral. On the other hand, the roots of the equation $x^2 - 3x + 2 = 0$ are contained in this set. It is possible to examine the set of all planets of the solar system, where we define *planets* as bodies moving around the sun in a closed orbit and weighing no less than one ton. The sun is not contained in this set, since it does not (strictly) move around itself. The earth is contained in this set. The Soviet rocket launched from the earth into an orbit about the sun on January 2, 1959, is also contained in this set; it is an artificial planet.

Let E be some set and N one of its elements. This relation is written symbolically as $N \in E$ and is read "N is an element of E." A symbolic

1. The question of what sort of method of determination is to be considered "effective" is of great interest in mathematical logic and philosophy, but it will not concern us here. An analogous difficulty is inherent in all formal classification systems. As an example, we may cite the biological difficulty in defining what sort of anthropomorphic beings belong to the class Homo sapiens.

11

notation of the type $E = \{L, M, N, \ldots\}$ is also used, where each element of the set is enumerated within the brackets. Thus, the set E_c, consisting of all of the capitals of the Soviet republics, could be written symbolically as $E_c = \{$Moscow, Kiev, Minsk, Tbilisi, Yerevan, Baku, Riga, Tallinn, Vilnius, Tashkent, Alma-Ata, Frunze, Ashkhabad, Dyushambe, Kishinev$\}$.

If every element of a set E is at the same time an element of a set E_1, the set E is called a *subset* of the set E_1. This is written as $E \subset E_1$ ("E is contained in E_1"). For example, the set of all integers is a subset of the set of all real numbers.

A set E is called *finite* if each of its elements can be associated with (mapped to) a different element of some set of the form $E_n = \{1, 2, 3, \ldots, n\}$. In other words, for a set E to be finite, there must exist a function F from E to E_n such that for each pair of elements a and b in E, $F(a) = F(b)$ implies $a = b$. For example, the set E_c of capitals of the Soviet republics is finite, since it is possible to enumerate this set using the elements of the set E_{15}, as is evident from table 3.1.

Table 3.1

Moscow	1	Yerevan	5	Vilnius	9	Ashkhabad	13
Kiev	2	Baku	6	Tashkent	10	Dyushambe	14
Minsk	3	Riga	7	Alma-Ata	11	Kishinev	15
Tbilisi	4	Tallinn	8	Frunze	12		

We are now in a position to give a definition of a *metric space*.

A metric space (E, d) is a set E in which for each pair of elements M and N a real number $d(M, N)$ is defined and the following properties are satisfied:

1. $d(M, N) = d(N, M)$ (symmetry).
2. $d(M, N) \geq 0$ (nonnegativity).
3. $d(M, N) = 0$ if and only if M and N are the same element (nondegeneracy).
4. $d(M, N) \leq d(M, L) + d(L, N)$ for each triple (M, N, L) of elements of the set E (triangle inequality).

We shall call the elements of the set E the *points* of the space (E, d). A metric space is thus completely determined by the choice of the set E and the function d—the *distance function* in the space. For the sake of simplicity, we shall denote a given space by the same letter as its corresponding set, although, in fact, the space and the set of its elements are quite different objects. In fact, it is often possible to define more than one distance function on a space E; each such function, along with the set E, determines a different metric space. In chapter 4 we shall construct new definitions of distance (and thus new metric spaces) in the plane.

In place of the four distance axioms listed above, it is possible to introduce only two (supposing as before that $d(M, N)$ is a real number):

1′. $d(M, N) = 0$ if and only if the points M and N are the same.

2′. $d(M, N) \le d(M, L) + d(N, L)$.

First of all, these properties follow from properties 1, 2, 3, and 4, as property 1′ is property 3, and property 2′ follows from the triangle inequality and condition 1.

On the other hand, from properties 1′ and 2′ alone it is possible to deduce all of the conditions 1, 2, 3, and 4.

To prove this, let us suppose first that in 2′, $L = M$, so that

$$d(M, N) \le d(M, M) + d(N, M).$$

By 1′, $d(M, M) = 0$. Therefore, $d(M, N) \le d(N, M)$. By interchanging in 2′ the positions of points M and N and carrying out the analogous argument, we see that $d(N, M) \le d(M, N)$. From these last two inequalities we get the axiom of symmetry (1):

$$d(M, N) = d(N, M).$$

Substituting M for N and N for L in 2′, we get

$$d(M, M) \le d(M, N) + d(M, N) = 2d(M, N),$$

so that, by virtue of 1′,

$$0 \le 2d(M, N),$$

implying

$$0 \le d(M, N),$$

which is property 2 (nonnegativity). Again, using the condition of symmetry which we proved above, we can interchange N and L in the second term on the right side of 2′ and get the triangle inequality 4. Thus, the system of axioms 1′ and 2′ is equivalent to the system 1, 2, 3, and 4. It is more convenient to use the latter system, however, as it gives in a clearer form the same fundamental properties of distance. Still, it is interesting to note that all of these properties can be embodied in a pair of axioms.

From the point of view of the definition which we have introduced, the content of the preceding chapter might be described as a proof that the set of points in three-dimensional Euclidean space along with a distance function defined as the length of the line segment joining a given pair of points is a metric space. In the end of the same chapter, we established that the set of points on the surface of a sphere, together with the distance function d_S, form a metric space.

We can get another example of a metric space if we consider the set of points of some surface π in three-dimensional space and define the distance $d_\pi(M, N)$ as the minimum length of the paths passing along the surface π and joining the points M and N.[2] The first three properties of distance are then immediately evident.

The triangle inequality can be verified in the following manner: Let us connect the points M and L, as well as the points L and N, by a path of the shortest possible length. Let us then connect the points M and N using such minimal paths ML and LN. Clearly, the length of this path cannot be less than the length of the shortest path joining M and N, since this path is itself a path joining M and N, and thus must be at least as long as the shortest such path. Since the length of this path is $d_\pi(M, L) + d_\pi(L, N)$, and the length of the shortest path between M and N is $d_\pi(M, N)$, the desired relation follows:

$$d_\pi(M, N) \leq d_\pi(M, L) + d_\pi(L, N) . \tag{3.1}$$

Let us note that on the surface of the sphere the shortest path joining two points is the smaller arc of the great circle determined by them; this was proved at the end of the preceding chapter. The proof was based on the fact that the triangle inequality was obtained by an independent argument concerned only with the space determined by the surface of a sphere, and on our proof that equality holds in the triangle inequality if and only if L lies "between" M and N on the smaller arc of a great circle.

It is useful to introduce the concept of *line segment* in an arbitrary metric space. We shall define the line segment joining the points M and N in a metric space E to be the set $E_{M,N}$ of points L which satisfy the equality

$$d(M, N) = d(M, L) + d(L, N) . \tag{3.2}$$

It is easy to see that for ordinary distance in the plane or in three-space, the set $E_{M,N}$ coincides with the line segment MN in the ordinary sense of

2. For the sake of simplicity, we suppose that for each pair of points M and N on a given surface π, there exists some shortest path between M and N. Using certain assumptions concerning the properties of the surface π, it is possible to prove this supposition.

the term. On the sphere S with the distance function d_S introduced in chapter 2, the segment $E_{M,N}$ is the smaller arc of the great circle joining the points M and N if M and N do not lie on the same diameter, and the whole sphere if the points M and N are antipodal.

We leave it for the reader to verify that with the distance $d_\pi(M, N)$ introduced above, the line segment $E_{M,N}$ (if it is indeed a unique path) is the shortest path (the so-called *geodesic line*) joining the points M and N.

It is also possible to generalize to an arbitrary metric space E the concept of the sphere $S_{M,r}$ with center M and radius r as the set of points N for which $d(M, N) = r$.

In the plane, this notion corresponds to that of a circle; in three-space, to that of the ordinary sphere; for the metric (distance function) d_S, to circles on the sphere S.

As still another (trivial) example of a metric space, we take an arbitrary set E and define the distance between two points M and N to be zero if they coincide, and one otherwise. It is easy to see that all of the necessary conditions are satisfied by this definition.

Various other examples of metric spaces will be examined in chapter 4.

In a metric space E it is always possible to define the concept of *convergence to a limit* for a sequence contained in E. Roughly speaking, a sequence of points in the metric space E, $(L_1, L_2, \ldots, L_k, \ldots)$, denoted by $(L_k)_{k=1}^{\infty}$, is said to *converge* to the point $L \in E$ if, beginning with some L_k, the distance between members of the sequence and the point L (the *limit*) becomes smaller than any previously chosen positive number.

Formally, the sequence $(L_k)_{k=1}^{\infty}$ is said to *converge* to L if for every positive real number ε it is possible to choose a positive integer $n(\varepsilon)$ such that the condition $k \geq n(\varepsilon)$ implies

$$d(L, L_k) < \varepsilon .$$

In keeping with the ordinary notation, we write

$$L = \lim_{k \to \infty} L_k .$$

It is easy to verify that for the metric space consisting of all real numbers $\mathbb{R}$ with a metric d defined by

$$d(x, y) = |x - y| ,$$

our general definition of limit coincides with the usual one.

For the metric space $\mathbb{R}^3$, Euclidean three-space with the usual metric, the concept of limit just defined allows us to state clearly what we mean by the limit of a sequence of points in three-space.

Let us note that in this case the set of points M for which $d(L, M) < \varepsilon$ forms the interior of a sphere with center L and radius ε. A sequence of points $(L_k)_{k=1}^{\infty}$ thus converges to the point L if and only if, for each $\varepsilon > 0$, there exists some integer $n(\varepsilon)$ such that all the points L_k of the sequence with $k \geq n(\varepsilon)$ lie in the interior of the sphere with center L and radius ε.

THEOREM. *If the sequence of elements* $L_1, L_2, \ldots, L_k, \ldots$ *of the metric space E converges to a limit L, then for each* $\varepsilon > 0$ *there exists an integer* $m(\varepsilon)$ *such that the conditions* $k \geq m(\varepsilon)$ *and* $k' \geq m(\varepsilon)$ *imply that* $d(L_k, L_{k'}) < \varepsilon$.

Proof. By the definition of limit, it is possible to choose an integer $n(\varepsilon/2)$ such that $k \geq n(\varepsilon/2)$ and $k' \geq n(\varepsilon/2)$ will imply the inequalities

$$d(L_k, L) < \frac{\varepsilon}{2}; \qquad d(L_{k'}, L) < \frac{\varepsilon}{2}.$$

But, by the triangle inequality and the axiom of symmetry,

$$d(L_k, L_{k'}) \leq d(L_k, L) + d(L, L_{k'}) = d(L_k, L) + d(L_{k'}, L) < \frac{\varepsilon}{2} + \frac{\varepsilon}{2} = \varepsilon.$$

In other words, if we let $m(\varepsilon) = n(\varepsilon/2)$, then for $k \geq m(\varepsilon)$ and $k' \geq m(\varepsilon)$, the following inequality holds:

$$d(L_k, L_{k'}) < \varepsilon.$$

This proves the theorem. To paraphrase slightly, we have proved that if elements of a sequence become arbitrarily "close" to a given limit, they also become arbitrarily "close" to each other.

If in the space E the converse of the above theorem holds, then E is called *complete*.

It is convenient to give the definition of a complete metric space in the following form: A sequence of points $(L_k)_{k=1}^{\infty}$ contained in the metric space E is said to be a *Cauchy sequence* if for each $\varepsilon > 0$, there exists an integer $m(\varepsilon)$ such that $k \geq m(\varepsilon)$ and $k' \geq m(\varepsilon)$ implies

$$d(L_k, L_{k'}) < \varepsilon.$$

The metric space E is called *complete* if each Cauchy sequence in E converges to a point of E.

The real line, the plane, and three-space with their usual metrics are complete metric spaces.

The question of whether or not a given metric space is complete is fundamental to the application of these concepts in mathematical analysis, but we shall not concern ourselves with this question at the present time.[3]

Two metric spaces are said to be *isometric* if it is possible to set up a one-to-one correspondence between them such that the distance between a pair of points in one of the spaces is the same as that between the corresponding points in the other space. From the point of view of the theory of metric spaces, two isometric spaces may be considered identical.

As an example, let the space E be the plane along with the ordinary metric, and the space E' the set of complex numbers z with a metric d' defined by the formula

$$d'(z, z_1) = |z - z_1| .$$

The usual method of picturing the complex numbers as points on the plane establishes the existence of a one-to-one correspondence between the two spaces. It is easy to check that this correspondence is an isometry, since, if we set $z = x + yi$ and $z_1 = x_1 + y_1 i$, the quantity

$$|z - z_1| = \sqrt{(x - x_1)^2 + (y - y_1)^2}$$

is equal to the distance between the corresponding points of the plane.

The definitions of metric space and of distance given here are not the most general encountered in mathematics. There are various generalizations of this concept. For instance, it would seem possible to assign infinite distance to some pairs of points, while still preserving all of the properties of distance. This generalization, as we shall see in chapter 9, is not particularly interesting. In many mathematical problems it is necessary to deal with a metric in which the property of symmetry is lacking. We shall study the properties of such a metric in chapter 9. In the theory of relativity, it is necessary to consider a distance function which can take on even imaginary values. The properties of such a distance are quite unique, but we shall not touch upon them in this book.

---

3. The notion of completeness is of most importance to mathematical analysis when applied to metric spaces whose points are functions. See, for example, the definition of the space C at the end of chapter 7.

4   Some Examples of
                                   Metric Spaces

In this chapter we shall look at a number of examples of metric spaces with relatively unusual metrics.

Many interesting metric spaces on the plane arise out of consideration of differently defined distance functions. We shall represent the points of the plane in this discussion with the aid of a coordinate system chosen once and for all so that each point of the plane is given by an ordered pair of coordinates (x, y). It will be convenient to denote a point of the plane as $M = (x, y)$.

The metric space l results when we define the distance between the points $M = (x, y)$ and $N = (x_1, y_1)$ by the formula

$$d_1(M, N) = |x - x_1| + |y - y_1| \,. \tag{4.1}$$

Figure 2.2 (p. 5) shows that $d_1(M, N)$ is the sum of the lengths of the legs of the triangle MLN, in which MN is the hypotenuse and the legs ML and LN are parallel to the axes of the coordinate system. Since the length of the hypotenuse cannot exceed the sum of the lengths of the legs, we have always

$$d(M, N) \le d_1(M, N) \,, \tag{4.2}$$

where $d(M, N)$ is the usual planar distance. The inequality (4.2) becomes an equality only when the line segment MN is either horizontal or vertical—that is, when it is parallel to one of the coordinate axes.

If in inequality (4.2) we substitute the algebraic expressions for the corresponding distance functions (4.1) and (2.2), we get the inequality

$$\sqrt{(x - x_1)^2 + (y - y_1)^2} \le |x - x_1| + |y - y_1| \,.$$

Setting $x_1 = y_1 = 0$, we get the simple but important inequality

$$\sqrt{x^2 + y^2} \le |x| + |y| \,. \tag{4.3}$$

18

Axioms 1, 2, and 3 are obviously satisfied by the metric $d_1(M, N)$. In order to verify that axiom 4 is also satisfied, we examine three points $M = (x, y)$, $N = (x_1, y_1)$, and $L = (x_2, y_2)$ and write the elementary identity

$$|x - x_1| + |y - y_1| = |x - x_2 + x_2 - x_1| + |y - y_2 + y_2 - y_1| . \tag{4.4}$$

Using the fact that for arbitrary real numbers a and b, $|a + b| \leq |a| + |b|$, from (4.4) we get the inequality

$$|x - x_1| + |y - y_1| \leq |x - x_2| + |x_2 - x_1| + |y - y_2| + |y_2 - y_1|,$$

which is the desired relation

$$d_1(M, N) \leq d_1(M, L) + d_1(L, N) . \tag{4.5}$$

And so the triangle inequality holds for the space l.

The distance $d_1(M, N)$ can be interpreted as the length of a minimal path traversed by a particle moving from M to N that is constrained to move only along line segments parallel to the coordinate axes. Figure 4.1 makes it evident that there are many (in fact, infinitely many) such minimal paths.

It is not hard to show that this statement is equivalent to saying that in the space l there exist infinitely many distinct *line segments*[1] joining the points M and N (except in the case where the points M and N are situated on the same vertical or horizontal line); for a line segment in the space l joining the points M and N is any broken line joining M and N which consists only of vertical and horizontal lines which do not intersect any vertical or horizontal line more than once. (We suggest that the reader prove this as an exercise.)

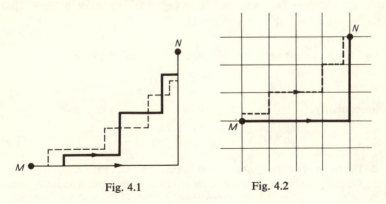

Fig. 4.1   Fig. 4.2

1. In the sense of the definition introduced in Chapter 3.

One gets a still more natural picture by considering the metric space C consisting of all the lattice points of some rectangular lattice in the plane (fig. 4.2) with the metric defined by formula (4.1). Points of this space can be viewed as the intersections of the streets of a perfectly planned city. The distance $d_1(M, N)$ is in this case the length of the shortest path which one can take along the streets of the city from the intersection M to the intersection N, without taking any shortcuts through houses.

In the following example, the space C will consist of points in the plane with the metric $d_\infty{}^2$ defined by the formula

$$d_\infty(M, N) = \max(|x - x_1|, |y - y_1|), \qquad (4.6)$$

where M has coordinates (x, y) and N has coordinates (x_1, y_1). Geometrically (fig. 2.2), the distance $d_\infty(M, N)$ can be interpreted as the length of the larger leg of the triangle MLN. As this length is always less than that of the hypotenuse (or equal to it in the case of a degenerate triangle), we have

$$d_\infty(M, N) \le d(M, N), \qquad (4.7)$$

where $d(M, N)$ is the usual planar distance. Again, setting $x_1 = y_1 = 0$, we get the algebraic inequality

$$\max(|x|, |y|) \le \sqrt{x^2 + y^2}. \qquad (4.8)$$

For the metric d_∞, axioms 1, 2, and 3 are again quite evident. To prove the triangle inequality, suppose we have three arbitrary points $M = (x, y)$, $N = (x_1, y_1)$, and $L = (x_2, y_2)$. We may assume that $|x - x_1| \ge |y - y_1|$.[3] This means that

$$d_\infty(M, N) = \max(|x - x_1|, |y - y_1|) = |x - x_1|$$
$$= |x - x_2 + x_2 - x_1|.$$

Consequently,

$$d_\infty(M, N) \le |x - x_2| + |x_2 - x_1|. \qquad (4.9)$$

2. The meaning of the symbol ∞ will be made clear on page 22.

3. We can make this assumption without loss of generality, for in the opposite case ($|x - x_1| < |y - y_1|$), we interchange the roles of the x and y coordinates and carry out the same proof.

Moreover, it is evident that

$$\left.\begin{aligned}|x - x_2| \le \max{(|x - x_2|, |y - y_2|)} = d_\infty(M, L)\,,\\ |x_2 - x_1| \le \max{(|x_2 - x_1|, |y_2 - y_1|)} = d_\infty(L, N)\,.\end{aligned}\right\} \quad (4.10)$$

Combining (4.9) and (4.10), we get

$$d_\infty(M, N) \le d_\infty(M, L) + d_\infty(L, N)\,, \qquad (4.11)$$

the desired result.

We have already noted that in an arbitrary metric space it is possible to introduce the concept of a sphere of radius r with center M, defined as the set of points N for which

$$d(M, N) = r\,. \qquad (4.12)$$

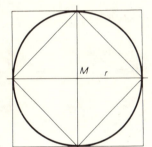

Fig. 4.3

If the distance function d is the ordinary distance on the plane, this sphere is just the circle with center M and radius r.

For three-space with the ordinary metric, the sphere defined by (4.12) is just the ordinary sphere with center M and radius r.

In the space l the sphere is a square with center M and diagonals of length $2r$ parallel to the coordinate axes.

In the space C the sphere is also a square with center M, but with sides of length $2r$ parallel to the co-ordinate axes. In figure 4.3 we have pictured the sphere of radius r in the spaces l and C and in the usual sense. The proof that spheres in l and C have the form indicated above is left as an exercise for the reader.

An interesting class of metric spaces is obtained when we define a metric d_p on the plane by the formula

$$d_p(M, N) = \sqrt[p]{|x - x_1|^p + |y - y_1|^p}\,. \qquad (4.13)$$

The spaces so obtained are called l_p *spaces.*

Axioms 1, 2, and 3 for a metric d_p are obvious. The triangle inequality follows from Minkowski's inequality:[4]

$$\sqrt[p]{|a + a_1|^p + |b + b_1|^p} \le \sqrt[p]{|a|^p + |b|^p} + \sqrt[p]{|a_1|^p + |b_1|^p}\,, \qquad (4.14)$$

4. A proof of Minkowsky's inequality can be found in Geoffrey H. Hardy, John E. Littlewood, and George Polya, *Inequalities* (Cambridge: The University Press, 1934).

which is true for $p \geq 1$, if for the points $M = (x, y)$, $N = (x_1, y_1)$, and $L = (x_2, y_2)$, we take

$$a = x - x_2; \qquad a_1 = x_2 - x_1; \qquad b = y - y_2; \qquad b_1 = y_2 - y_1.$$

For $p < 1$, the triangle inequality is not true; the inequality in (4.14) is reversed.

It is easy to see that for $p = 1$ the distance $d_p(M, N) = d_1(M, N)$, whereas for $p = 2$ the distance $d_p(M, N)$ is just the usual distance $d(M, N)$. Thus, the space l coincides with the space l_1, and the plane with the usual metric is the space l_2.

We shall now show that the distance $d_p(M, N)$ converges to the distance $d_\infty(M, N)$ as $p \to \infty$.

Let us first examine the case $|x - x_1| > |y - y_1|$. Then $d_\infty(M, N) = |x - x_1|$. On the other hand, transforming (4.13), we have

$$d_p(M, N) = |x - x_1| \sqrt[p]{1 + \left| \frac{y - y_1}{x - x_1} \right|^p}. \tag{4.15}$$

Noticing that for $p > 1$,

$$1 \leq \sqrt[p]{1 + \left| \frac{y - y_1}{x - x_1} \right|^p} \leq 1 + \left| \frac{y - y_1}{x - x_1} \right|^p$$

and that the quantity $|(y - y_1)/(x - x_1)|^p \to 0$ as $p \to \infty$ (since $|x - x_1| > |y - y_1|$ and, thus, $|y - y_1|/|x - x_1| = |(y - y_1)/(x - x_1)| < 1$), we get

$$\lim_{p \to \infty} \sqrt[p]{1 + \left| \frac{y - y_1}{x - x_1} \right|^p} = 1.$$

Using (4.15), we see that

$$\lim_{p \to \infty} d_p(M, N) = |x - x_1| = d_\infty(M, N). \tag{4.16}$$

Analogously, for $|x - x_1| < |y - y_1|$, we obtain

$$\lim_{p \to \infty} d_p(M, N) = |y - y_1| = d_\infty(M, N). \tag{4.17}$$

Finally, let us examine the case $|x - x_1| = |y - y_1|$. Then

$$d_p(M, N) = \sqrt[p]{2|x - x_1|^p} = |x - x_1|\sqrt[p]{2} \,.$$

Since $\lim\limits_{p \to \infty} \sqrt[p]{2} = 1$, we have in this case

$$\lim_{p \to \infty} d_p(M, N) = |x - x_1| = d_\infty(M, N) \,. \tag{4.18}$$

And so in all three cases, by (4.16), (4.17), and (4.18), we get

$$\lim_{p \to \infty} d_p(M, N) = d_\infty(M, N) \,, \tag{4.19}$$

the desired result.

Consequently, it is reasonable to denote the space C by the symbol l_∞, since the distance $d_\infty(M, N)$ in this space is the limit of the distances $d_p(M, N)$ as p approaches infinity.

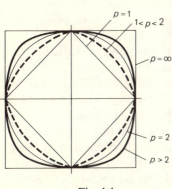

Fig. 4.4

Figure 4.4 depicts l_p spheres (all having the same center M) for various values of p. The l_p metric spaces are also called *Minkowski spaces*. In chapter 7 we shall examine multidimensional Minkowski spaces.

We leave it to the reader to formulate a simple definition of *line segment* for l_p spaces.

We can obtain an interesting class of metric spaces in the plane by defining distance as the minimum time required to travel from M to N with some given restrictions on the paths which may be taken.

Making no restrictions, we can obtain the usual distance if the shortest path from M to N is taken by a point moving with a constant velocity of one.

We can obtain the metric space l if we require this point to move again with constant velocity but only along line segments parallel to the coordinate axes.

But we get a new example if (see fig. 4.5) we consider the map of the Moscow metropolitan area and suppose that a traveler may go from point M to point N in the following manner.

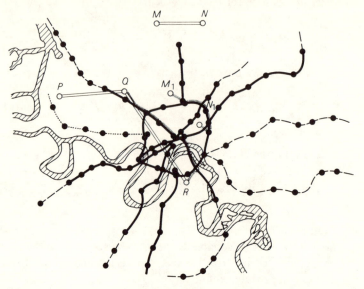

Fig. 4.5

If the same subway station is the nearest one to both points, the shortest route is on foot. If this is not the case, the traveler walks (by the shortest route) to the station closest to the point of departure M, rides by the shortest route to the station closest to the point N, and from there walks to N. If two or more subway stations are equally close to M or N, the route for which the riding time will be least is chosen. Figure 4.5 shows two pairs of points, (M, N) and (M_1, N_1); to go from M to N one must walk, whereas to go from M_1 to N_1 one must take a subway. Let us suppose that someone living between the Rizhskii and Botanic stations wants to go somewhere in the neighborhood of the Zemlyanii Val; then it would be necessary to get on at the Botanic station and go to either the Lermontov station or the Kursk station. It is easy to see that the metric $d_t(M, N)$ defined in this way is, in general, different from the usual geometric distance. In fact, if the point Q is situated near a heloport (either Dynamo or Aeroport), the point P near Volokolamskii Highway, and the point R near Valovaya Street (near the Paveletskii subway station) as in figure 4.5, then in the sense of ordinary distance the point P is somewhat closer to the point Q than is R:

$$d(P, Q) < d(Q, R).$$

It is evident from figure 4.5, however, that

$$d_t(P, Q) > d_t(Q, R).$$

Actually, if one cannot take a taxi, it is possible to travel from the heloport to Valovaya Street in less time than it takes to go from the heloport to Volokolamskii Highway.

For the metric d_t, axiom 1 (the axiom of symmetry) is nontrivial. The equality

$$d_t(M, N) = d_t(N, M)$$

indicates that the time spent in going from M to N as quickly as possible is the same as that spent going from N to M. This is more or less true if one uses only the subway or travels only on foot. But if taxis are allowed, this is no longer true; it is one thing to try to get a taxi at a taxi stand and an entirely different thing to try to get one in some remote neighborhood or at the Kursk station when the trains are coming in.

Axioms 2 and 3 for the metric d_t are evident. The reader will have no trouble proving the triangle inequality (axiom 4) for himself if he recalls the proof of this axiom for the metric d_π in chapter 3.

In further investigation into the properties of metric spaces, it will be useful to introduce the concept of a *Dirichlet region*. Let E be a metric space and $L_1, L_2, \ldots, L_k$ points in E. We define the *Dirichlet region of the point L_i* to be the set of all points N for which

$$d(L_i, N) \leq d(L_j, N) \tag{4.20}$$

for all $j \neq i$, and denote this set by D_i. In other words, the Dirichlet region D_i is the set of points which are at least as close to the point L_i as to any of the other given points L_j. It is clear that the Dirichlet region is determined by the choice of the points $L_1, L_2, \ldots, L_k$ and of the given point L_i. We shall now look at examples of Dirichlet regions in various metric spaces.

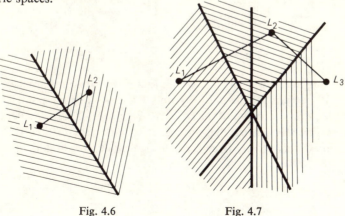

Fig. 4.6   Fig. 4.7

Let us first consider the plane with the usual metric d_2 and two points L_1 and L_2. We join these points by the line segment L_1L_2 (fig. 4.6) and draw a perpendicular through its midpoint. This perpendicular divides the plane into two closed half-planes, which are the Dirichlet regions for the points L_1 and L_2.

Let us now consider three points L_1, L_2, and L_3 in the plane, again with the usual metric. In figure 4.7 we have constructed the Dirichlet regions for these three points and marked them off with heavy lines. The method of construction is clear from the diagram.

Let us now examine two points L_1 and L_2 in the plane with the metric $d_1(M, N)$ (that is, in the space l). For the sake of clarity, we shall again visualize a city divided into squares. The Dirichlet regions consist of those intersections from which the route through the city to L_1 will be shorter than that to L_2 and vice versa. These regions are marked off in figure 4.8 by a heavy line. Figure 4.9 shows the corresponding partition for the space C. We suggest that the reader try to derive the general rule for constructing the Dirichlet regions for n points in the spaces l and C by examining Dirichlet regions for two points and for three points.

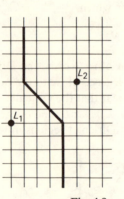

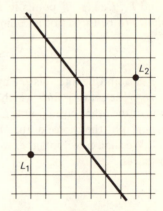

Fig. 4.8   Fig. 4.9

Turning again to figure 4.5, we see that if we partition the space into Dirichlet regions for the pair of points P and R, then the point Q falls into the Dirichlet region of the point R. We suggest that the reader draw this partition into Dirichlet regions. It is important to note that this partition differs greatly from those in figures 4.6, 4.8, and 4.9.

5   The Space of Information

When we speak of *communication*, we usually mean some sort of transmission of information. In this sense, communication appears in the form of books, letters, telegrams, musical pieces (recorded or written in musical notation), computer cards, signals directing the flight and landing of space ships, molecules of deoxyribonucleic acid (DNA) which transmit genetic information from parents to offspring, and so on.

Questions concerning the transmission and codification of information are examined in the theory of information.[1] In the study of this theory, methods for determining the "quantity of information" contained in a given message are developed; this "quantity" can itself be encoded as information. We frequently encounter this situation in our daily lives; in composing a telegram we try to use the minimum number of words possible without destroying the meaning (that is, while preserving the quantity of information).

The reverse situation arises when, in an examination or in a seminar, a poorly prepared student amplifies his message, trying to express the small amount of information which he has on his topic in a sufficiently impressive quantity of words.

A surplus of communication relative to the quantity of information to be transmitted is, however, not always harmful. Such redundancy can be useful when interference arises in the transmission of information.

For example, when we have a bad connection on the telephone, we are forced to repeat individual words. In conveying strange or difficult

1. A good reference for an account of information theory is A. M. Yaglom and I. M. Yaglom, *Veroyatnost' i informatsiya* [Probability and Information] (Moscow: State Publishing House of Physics and Mathematics Literature, 1960). A translation of this work will be included in the Survey of Recent East European Mathematical Literature of the University of Chicago.

names, we use the following alphabetical device: In communicating the name "Pavsikakii" over the telephone, for example, we might say, "Peter, Anne, Victor, Susan, Irene, Karen, Albert, Kay, Ivan, Ida."

In this chapter and in the next we shall study methods of *error-stabilizing codification* of information, without concerning ourselves with specific questions relating to the theory of information. In other words, we shall study methods of writing down messages that allow us to correct automatically any errors that arise, provided that they are not too numerous. These methods are closely connected with the question of the possibility of defining a metric on the so-called space of information.

The idea of these methods is something we make use of frequently in everyday life—for instance, in reading books with printing errors and receiving telegrams with mistakes. If we read the word "sauce pin" in a book, we need not look in any "dictionary of mistakes" in order to guess its meaning. There is very little chance that the author meant the word "telegraph" here. For if he did, we would be dealing with eight misprints in a row, whereas if the word "sauce pan" was meant, there would be only one misprint.[2]

Still, there are curious examples where a totally different word can arise from a mistake in only one letter. For example, the Russian word "korona" ("crown") could be mistakenly written as "korova" ("cow") or as "vorona" ("crow").

Indeed, a well-known anecdote is based on this situation. A Russian provincial newspaper is said to have printed this sentence in an article about the coronation of Nicholas II: "The Metropolitan placed the crow on His Highness's head." The next day a correction was published: "The Metropolitan placed the cow on His Highness's head."

Clearly, even here it is quite easy to determine the true meaning of the message from the context.

Analogously, a misprint in a musical composition can frequently be discovered because of its false sound and can be corrected by the laws of harmony.

One must mention that errors can arise not only in transmission of information, but also during its storage, for example, in the memory of an electronic computer. The problem of discovering the correct message is the same for errors occurring during the storage of the message as for those arising during transmission.

Every type of message is written with the aid of some set of symbols. The set of symbols used forms an *alphabet* $\mathfrak{A}$. We assume that this

---

2. Of course, sometimes there are more probable strings of misprints, arising from a typist's or typesetter's misunderstanding of the sense of certain words.

alphabet is given beforehand and consists of a finite number of symbols. For example, the alphabet might consist of all Russian letters, a space, and punctuation marks. Using this alphabet, it is possible to write any arbitrary Russian sentence. Another example of an alphabet is the set of all decimal digits, algebraic symbols, punctuation marks, and Latin and Greek letters. Using such an alphabet, one can write down the most diverse of mathematical formulas.

Still another example is the binary alphabet—a set of two symbols, $\mathfrak{A}_2 = \{0, 1\}$. Using such an alphabet, we can write any number in the binary system.

It is easily verified that any whole number x can be written in the form

$$x = \varepsilon_n 2^n + \varepsilon_{n-1} 2^{n-1} + \cdots + \varepsilon_1 2 + \varepsilon_0 , \qquad (5.1)$$

where the quantities ε_i take on a value of 0 or 1.

Thus, to transmit information about an integer x, it suffices to transmit a finite sequence of symbols of the alphabet $\mathfrak{A}_2$: $\varepsilon_n, \varepsilon_{n-1}, \ldots, \varepsilon_1, \varepsilon_0$. In order to separate the information about two different numbers, it is necessary either to introduce a special symbol for the end of a number or to transmit only sequences of some standard length.[3]

The latter method is the one actually used on computers, where the binary sequences to be stored in memory usually have a standard length corresponding to the number of "memory cells" available in the machine. In computers now being manufactured, however, this principle is being departed from more and more, with memories of variable length being used.

Formula (5.1) is analogous to the well-known formula

$$x = a_n 10^n + a_{n-1} 10^{n-1} + \cdots + a_1 10 + a_0 , \qquad (5.2)$$

where $a_n, a_{n-1}, \ldots, a_1, a_0$ are the digits in the decimal representation of the number x. It is easy to generalize equality (5.1) to numbers which are not integers exactly as is done for decimal fractions.

Let us determine the connection between the number n and the value of x in (5.1). Clearly, if the leading coefficient is equal to zero, the leading term can be discarded; this process can be carried out repeatedly until

3. There are more complicated methods for separating the meaningful units (words) in an arbitrary alphabet. See, for example, the article by A. A. Sardinas and George Patterson in *Kiberneticheskii sbornik* [Journal of Cybernetics], no. 3, Moscow, 1961.

$\varepsilon_n = 1$. Switching all terms except the first, from the right side to the left, we get

$$x - \varepsilon_{n-1}2^{n-1} - \varepsilon_{n-2}2^{n-2} \cdots - \varepsilon_1 2^1 - \varepsilon_0 = \varepsilon_n 2^n = 2^n .$$

This makes it clear that

$$2^n \leq x$$

or

$$n \leq \log_2 x . \qquad (5.3)$$

On the other hand, the following inequality holds:

$$\varepsilon_n 2^n + \varepsilon_{n-1}2^{n-1} + \varepsilon_{n-2}2^{n-2} + \cdots + \varepsilon_1 2^1 + \varepsilon_0$$
$$\leq 2^n + 2^{n-1} + 2^{n-2} + \cdots + 2^1 + 1 = 2^{n+1} - 1 .$$

From this relation and from (5.1), it follows that

$$x \leq 2^{n+1} - 1$$

or

$$x < 2^{n+1} ,$$

which can also be written

$$n + 1 > \log_2 x . \qquad (5.4)$$

Combining (5.3) and (5.4), we obtain the inequality

$$n \leq \log_2 x < n + 1 . \qquad (5.5)$$

The inequality (5.5) can be written as

$$n = [\log_2 x] ,$$

that is, n is equal to the *greatest integer* in $\log_2 x$.[4] The above statement leads us to conclude that the number of binary symbols required to code all integers in the range $0 \leq x \leq a$ is

$$1 + [\log_2 a] = 1 + n . \qquad (5.6)$$

4. By the *greatest integer in the number* a we mean the largest integer which is less than or equal to a. The greatest integer in a is written $[a]$; for example, $[\pi] = 3$.

The one is included here because there are $n + 1$ terms in the sequence $\varepsilon_n, \varepsilon_{n-1}, \ldots, \varepsilon_1, \varepsilon_0$.

With the aid of the binary alphabet $\mathfrak{A}_2$, any type of information (numbers, commands, logical relations, and so forth) can be written into the memories of computers.[5]

By a *message* in a given alphabet $\mathfrak{A}$ we shall mean a finite sequence of symbols from this alphabet. It is sometimes convenient to divide a message into standard submessages, which are called *words*.

Generally speaking, it is possible to define infinite alphabets and messages, but we shall not consider them here.

A message written in one alphabet can sometimes be translated into another. For example, as we have already seen, an integer represented by its decimal digits can also be written in the binary alphabet. One of the important examples of such translation is the following: Suppose that we are given an alphabet $\mathfrak{A}$. We define a new alphabet $\mathfrak{A}'$ to be the set of all words of length less than or equal to some positive integer k which can be formed using alphabet $\mathfrak{A}$. It is clear that every message in alphabet $\mathfrak{A}$ can be broken up into a sequence of words of length not greater than k, which means that it can be recoded in the new alphabet $\mathfrak{A}'$.

A similar idea could be introduced for messages in the Russian language, written in the Russian alphabet supplemented by a space and punctuation marks. Here it would be necessary to take a complete word list of the Russian language and to assign to each word in it a hieroglyph (using, for example, a combination of Chinese and Egyptian writing). If one could, in addition, introduce hieroglyphs in such a way that it would be possible to distinguish cases and conjugations of verbs, then one could recode any message in the Russian language.

In place of hieroglyphs one might use decimal numbers of six digits. The first five digits of such numbers would suffice for coding words;[6] the sixth digit could be used for coding grammatical signs.

Here we have for the first time stumbled upon the important notion of coding and recoding messages. By *codification* we mean, generally speaking, the formation in a given alphabet of messages containing given information or the translation of a message written in one alphabet into a message written in another. In this respect, "one-to-one" translations, that is, cases in which it is possible to transform the information of a message from one language to another in an essentially unique way, are of most interest. It is easy to see that the translation of

5. On this point, see Donald E. Knuth, *The Art of Computer Programming*, vol. 1 (Reading, Mass.: Addison-Wesley, 1969).

6. As one could easily make do with a vocabulary of 100,000 Russian words.

Russian sentences from the alphabetical to the hieroglyphic form has this property.

In practice, this method of encoding messages by words is used along with a method of decoding by means of a *word alphabet*.

The reverse situation also occurs, in which a symbol from a given alphabet 𝔄 is coded in the form of a word written in a simpler alphabet 𝔄'. For example, suppose an alphabet consists of three symbols $\{\cdot, -, *\}$ (dot, dash, end of letter). Then an arbitrary letter or punctuation mark can be written in this *Morse code* (see table 5.1) as a word of at most seven symbols from the alphabet 𝔄'.

Table 5.1
The Morse Alphabet

Morse Symbols	Latin Letters	Morse Symbols	Latin Letters (and Arabic Numerals)
· —	A	· · · —	V
— · · ·	B	· — —	W
— · — ·	C	— · · —	X
— · ·	D	— · — —	Y
·	E	— — · ·	Z
· · — ·	F	· — — — —	1
— — ·	G	· · — — —	2
· · · ·	H	· · · — —	3
· ·	I	· · · · —	4
· — — —	J	· · · · ·	5
— · —	K	— · · · ·	6
· — · ·	L	— — · · ·	7
— —	M	— — — · ·	8
— ·	N	— — — — ·	9
— — —	O	— — — — —	0
· — — ·	P	· — · · —	, (comma)
— — · —	Q	· · · · · ·	. (period)
· — ·	R	— · — · — ·	; (semicolon)
· · ·	S	— — — · · ·	: (colon)
—	T	· · — — · ·	? (question mark)
· · —	U	— — · · — —	! (exclamation point)

The marks "*" and "**" for the end of a letter and the end of a word, respectively, are coded by intervals of time and, therefore, are not included in the table.

In this way, English words can be written in the Morse alphabet instead of the Latin alphabet.

Example. The English sentence, "What is distance?" can be written as follows in the Morse alphabet:

. — — *. . . . *. — * — **. . *. . . **

— . . *. . *. . . * — *. — * — . * — . — . *. *. . — — . . **

As it happens (fortunately for computer technology), any message in an arbitrary finite alphabet can be recoded in the binary alphabet $\mathfrak{A}_2 = \{0, 1\}$.

Any nonnegative integer can be represented in the form of equation (5.1); that is, in the binary system and, therefore, as a word in the binary alphabet.

If we consider only integers in some range $0 \le x \le a$, the sequence of binary symbols for $x \sim \varepsilon_n \varepsilon_{n-1} \cdots \varepsilon_1 \varepsilon_0$ can have no more than $1 + [\log_2 a]$ terms (5.6).

Now, if we have an arbitrary finite alphabet $\mathfrak{A}$ consisting of m symbols, we can assign to each symbol an integer between 0 and $m - 1$ inclusive. And so, to each symbol of the alphabet $\mathfrak{A}$ it is possible to assign a binary word, corresponding, in accordance with (5.1), to the number associated with that symbol. Moreover, it is possible to make do with words of length n, where

$$\log_2 (m - 1) < n \le 1 + \log_2 (m - 1). \tag{5.7}$$

In this way it is possible in the case of any finite alphabet to limit oneself to words in the binary alphabet. Modern telegraphy employs an international telegraphic code for Russian and Latin letters, numerals, and punctuation marks. As an example, we introduce in table 5.2 the five-symbol code used in telegraphic apparatus of type CT-35.[7]

The last five combinations are read in the same way in all registers.

The symbols of the registers indicate that after the appearance of, let us say, the symbols of the Latin register, all binary five-symbol combinations are read as Latin letters. In order to switch to Russian letters, one must insert the symbol for the Russian register.

Example. Let us write the following sentence in our telegraph code: "The name *Shakespeare* is written *Шекспир* in Russian."

7. At present the so-called "international telegraphic code No. 2" is being used more and more. The following code, a variation of the "international telegraph alphabet No.1" for multiplex systems, is based on an analogous principle.

00001 10101 11010 01000 10001 01111 10000 01011 01000
10001 00101 11010 10000 10011 01000 00101 11000 01000
10000 00111 01000 10001 01100 00101 10001 01101 00111
01100 10101 10101 01000 01111 10001 **00010** 01100 **11111**
01000 10011 00101 11000 01100 00111 10001 00001 01100
01111 10001 00111 10100 00101 00101 01100 10000 01111
00010 00101

Table 5.2
International Telegraphic Code for Russian and
Latin Letters

Latin Register	Numerical Register	Russian Register	Code Combination
A	1	А	10000
B	8	Б	00110
W	?	В	01101
G	7	Г	01010
D	0	Д	11110
E	2	Е	01000
V	'	Ж	11101
Z	:	З	11001
I	Ш	И	01100
J	6	Й	10010
K	(	К	10011
L	=	Л	11011
M	)	М	01011
N	Ю	Н	01111
O	5	О	11100
P		П	11000
R	—	Р	00111
S	.	С	00101
T		Т	10101
U	4	У	10100
F	Э	Ф	01110
H	+	Х	11010
C	9	Ц	10110
Q	/	Щ	10111
X	,	Ъ	01001
Y	3	Ы	00100
		Я	00011
Russian Register			11111
Numerical Register			00010
Latin Register			00001
blank			10001
bell			00000

In the coded text the symbols for the registers are in heavy print, for it is necessary to notice them in order to be able to change from Russian to English, to arrange the punctuation marks, and to write down the letter "ш." The latter is placed in the numerical register since there are more Russian than Latin letters.

A five-digit binary code suffices for the representation of all Latin (or Russian) letters. Such a code is given in table 5.3.

Table 5.3

a 00000	h 00111	o 01110	u 10100
b 00001	i 01000	p 01111	v 10101
c 00010	j 01001	q 10000	w 10110
d 00011	k 01010	r 10001	x 10111
e 00100	l 01011	s 10010	y 11000
f 00101	m 01100	t 10011	z 11001
g 00110	n 01101		

The sentence "The length of the hypotenuse is less than the sum of the lengths of the two legs" can be coded as follows:

10011	00111	00100	11010	01011	00100	01101	00110	10011
00111	11010	01110	00101	11010	10011	00111	00100	11010
00111	11000	01111	01110	10011	00100	01101	10100	10010
00100	11010	01000	10010	11010	01011	00100	10010	10010
11010	10011	00111	00000	01101	11010	10011	00111	00100
11010	10010	10100	01100	11010	01110	00101	11010	10011
00111	00100	11010	01011	00100	01101	00110	10011	00111
10010	11010	01110	00101	11010	10011	00111	00100	11010
10011	10110	01110	11010	01011	00100	00110	10010	

Note that the separation of the five-digit strings is used here only for ease of reading and that the blank entry 11010 has been introduced as a space symbol between words. For storage of such a message in the memory of a computer or for transmission by means of telegraph, no symbols but zero and one are needed.

To illustrate this point, let us suppose that the above text were written as a continuous string of zeros and ones. Then the first line (excluding space symbols) would read:

$$10011001110010001011001000110100110100 11$$

We could initially separate the first five symbols 10011 and write them down. Then we could separate the immediately following five symbols 00111 and write them down. In this way we could generate the complete message by inserting spaces between strings (on the subject of separating words in messages, see note 3 on page 29).

We shall now introduce the idea of a *space of communication*. Let us consider an arbitrary alphabet[8] $\mathfrak{A}$ and the set of messages consisting of exactly n symbols from the alphabet $\mathfrak{A}$.

We define the distance $d(\xi, \eta)$ between two messages ξ and η to be the number of positions in which the messages ξ and η have different symbols. The metric space $E(n, \mathfrak{A})$ obtained in this way is called the n-dimensional space of communication over the alphabet $\mathfrak{A}$.

Example 1. $\mathfrak{A}$ is the Latin alphabet, $n = 5$. Let $\xi = \text{build}$; $\eta = \text{guilt}$. All letters but the first and fifth coincide, and so $d(\xi, \eta) = 2$.

Example 2. $\mathfrak{A}_2$ is the binary alphabet, $n = 12$ and $\boldsymbol{\xi} = 000110101010$; $\eta = 010101101011$. The second, fifth, sixth, and twelfth binary digits do not coincide, and so $d(\xi, \eta) = 4$.

Note that it is possible to compare any words of length not greater than n if it is agreed that words of less than n letters are augmented by a previously chosen letter until they are of length n (usually 0 in the binary alphabet).

Let us verify that the metric d defined above satisfies all of the necessary axioms.

The axiom of symmetry $d(\xi, \eta) = d(\eta, \xi)$ follows from the definition, in which the roles of ξ and η are interchangeable. It is obvious that $d(\xi, \eta) \geq 0$, and that $d(\xi, \eta) = 0$ only if all of the corresponding symbols in the messages ξ and η coincide—that is, if the words ξ and η are the same.

The triangle inequality is verified as follows: Assume that we are given three words ξ, η, and ζ of length n. Let us suppose that in the kth position, the symbols of words ξ and ζ coincide, as well as the symbols of words ζ and η. Then it is clear that for this position the symbols of words ξ and η also coincide.

To be concise, let ξ_k be the kth symbol of message ξ; ζ_k the kth symbol of message ζ; and η_k the kth symbol of message η. Then if $\xi_k = \zeta_k$ and $\zeta_k = \eta_k$, $\xi_k = \eta_k$. Taking the contrapositive, if $\xi_k \neq \eta_k$, then either $\xi_k \neq \zeta_k$ or $\zeta_k \neq \eta_k$.

Thus, words ξ and η can have different symbols only in those positions where either the symbols of words ξ and ζ or those of words ζ and η do not coincide. This indicates that the number of symbols of ξ and η which differ does not exceed the sum of the number of noncoincident

8. In this situation it is not even essential that the alphabet $\mathfrak{A}$ be finite.

symbols of ξ and ζ and those of ζ and η. But the number of symbols at which ξ and ζ do not coincide plus the number at which ζ and η do not coincide is $d(\xi, \zeta) + d(\zeta, \eta)$. In other words,

$$d(\xi, \eta) \leq d(\xi, \zeta) + d(\zeta, \eta), \qquad (5.8)$$

the triangle inequality.

Example. In the space $E(5, \mathfrak{A})$, where $\mathfrak{A}$ is the Latin alphabet, let ξ = trace, η = truce, and ζ = trunk. Clearly, $d(\xi, \eta) = 1$, $d(\xi, \zeta) = 3$, and $d(\zeta, \eta) = 2$; and so

$$d(\xi, \eta) \leq d(\xi, \zeta) + d(\zeta, \eta).$$

With the aid of this metric, it is possible to formulate a general principle for the construction of codes which allows us to correct mistakes automatically. This principle was first introduced by P. Hemming.[9] We shall examine it in the following chapter.

9. See the article by P. Hemming in the collected translations *Kody s obnaruzheniem i ispravleniem oshibok* (*Codes and the Detection and Correction of Errors*), IL, Moscow, 1956.

6 Automatic Correction of Errors in Messages

In this chapter we shall examine the space of communication $E(n, \mathfrak{A})$; that is, the space of messages of length n formed in the alphabet $\mathfrak{A}$. As we have already seen, it is possible to limit oneself exclusively to binary messages (messages in the alphabet $\mathfrak{A}_2$). All of our interesting examples will come from this alphabet.

Let us consider the following general scheme for the transmission of information (fig. 6.1). Messages emanating from some source are recoded into an error-stabilizing code by means of a coding device. Then these messages are transmitted along connecting lines, during which time the messages might become distorted. Finally, the messages are corrected at the receiving end by a decoding device and decoded into the initial code if necessary.

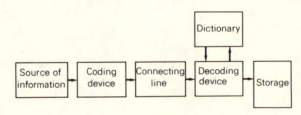

Fig. 6.1

The automatic detection and correction of errors during the storage of information in machine memory occurs in a completely analogous manner.

As information is stored in the memory, it is translated into an error-stabilizing code. When the message is read, the corresponding decoding takes place, along with the correction of errors admitted during storage. By periodically reading, decoding, correcting, recoding, and storing

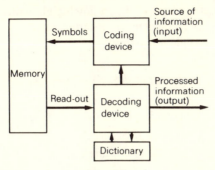

Fig. 6.2

information, we can be sure of its accuracy. In particular, if we choose a period of time T during which not too many distortions of the stored information can arise, and repeat the above process no less frequently than once per period T of time, the accuracy of the stored information will be guaranteed. In other words, T must be chosen so small that the distance $d(\xi, \xi')$ between the stored message ξ and the message ξ' that is read cannot become too great.

We now choose a subset N_k of $E(n, \mathfrak{A})$ with the property that for any two distinct elements ξ and η of N_k,

$$d(\xi, \eta) \geq k . \tag{6.1}$$

We shall call the set N_k the *set of intelligible words*. Let us suppose that during the transmission and storage of the intelligible word $\xi \in N_k$, l errors (with $l \leq k - 1$) are admitted—that is, l symbols of the word ξ are incorrectly given. This incorrectly transmitted word we shall denote as ξ'. By the definition of our metric, $d(\xi, \xi') = l$. Clearly, the word ξ' is not intelligible, for if it were, $d(\xi, \xi')$ would be greater than l—by (6.1).

Thus, we may check the transmitted word ξ' and see that it is not intelligible (it can, for example, be compared with all the words of N_k—in figures 6.1 and 6.2 this possibility is guaranteed by the availability of a dictionary). We would then discover that an error had been made. While the word is in the machine memory, such a process of checking can take place periodically, where we choose the period T to satisfy the condition that during the time T there is little chance for more than $k - 1$ errors to arise in a single word. Thus, we already have a general principle for the detection of errors.

But we can do even more; we can actually correct the mistakes that arise. For this purpose we shall assume that the number of mistakes $l \leq (k - 1)/2$. Let η be an arbitrary intelligible word distinct from ξ, and ξ' as before, an incorrectly transmitted word.

Applying the triangle inequality,

$$d(\xi, \eta) \leq d(\xi, \xi') + d(\xi', \eta).$$

Setting $d(\xi, \xi') = l$, and using (6.1),

$$k \leq l + d(\xi', \eta).$$

From this it is clear that

$$d(\xi', \eta) \geq k - l \geq k - \frac{k - 1}{2} = \frac{k + 1}{2}, \qquad (6.2)$$

since $l \leq (k - 1)/2$.

From the assumption that $l \leq (k - 1)/2$ and from (6.2), we conclude that the incorrect word ξ' is at most $(k - 1)/2$ away from the correct word ξ and at least as far as $(k + 1)/2$ from any other intelligible word η. In other words, we find that the intelligible word ξ is nearest to ξ', and thus establish the correct message.

The above discussion seems to suggest the usefulness of determining the Dirichlet region of each intelligible word. For each word ξ' to be corrected, it must be determined to which Dirichlet region the message belongs. The intelligible word determining this Dirichlet region is considered the correct word.

Here is where Hemming's remarkable idea comes in. This idea is based on the fact that for the purposes of transmission, one need not use all possible combinations of symbols from the alphabet, but only some set of intelligible words. Since in English only certain combinations of letters are used as intelligible words, the sense of distorted words can frequently be established without the use of additional codings. This has already been illustrated.

We shall now examine the means by which error-stabilizing codes are constructed in practice; in particular, the construction of sets of intelligible words $N_k \subset E(n, \mathfrak{A})$. All of our examples will come from the binary alphabet $\mathfrak{A}_2$. As we have already seen, such a condition is not a limitation, for it is possible to write any message in the binary alphabet.

The problem of error-stabilizing codification can be formulated in the following way. Suppose that we have the space of s-symbol binary messages $E(s, \mathfrak{A}_2)$. We must place in correspondence with each such message a message from some set $N_k \subset E(n, \mathfrak{A}_2)$. This set of intelligible words N_k must be *stable* with respect to l-place errors. We shall call the quantity $(n - s)/n$ the *redundancy* of the code.

Since the exact formulation of this problem must involve the probable distortion occurring during transmission, it is necessary to construct the code (the set $N_k \subset E(n, \mathfrak{A}_2)$) in such a way that the probability of receiving more than l errors in a word of length n is sufficiently small. This more complete formulation of the problem is studied in information theory, but it need not concern us here.

In the construction of these codes it is especially convenient to introduce and apply the concept of addition *modulo 2*; that is, according to the rules

$$0 \oplus 0 = 0, \qquad 1 \oplus 0 = 1,$$
$$0 \oplus 1 = 1, \qquad 1 \oplus 1 = 0.$$

The circled plus sign indicates that the operation carried out is not ordinary addition. The distance between two binary words $\xi = x_1 x_2 \cdots x_n$ and $\eta = y_1 y_2 \cdots y_n$ (where all x_i and y_i have the value 0 or 1) can now be written in the following way:

$$d(\xi, \eta) = (x_1 \oplus y_1) + (x_2 \oplus y_2) + \cdots + (x_n \oplus y_n).$$

Since ones will appear as terms in this sum when and only when corresponding symbols differ, the total will be exactly equal to the defined distance $d(\xi, \eta)$.

Let us consider the space of messages $E(n, \mathfrak{A}_2)$ and associate with each word $\xi \in E(s, \mathfrak{A}_2)$ a word ξ' of length $s + 1$, formed according to the following rule: The first s symbols of the word ξ' will be the same as those of the word ξ. The last $((s + 1)^{\text{st}})$ symbol of the word ξ' is chosen in such a way as to make the sum (ordinary) of binary symbols in the word ξ' even. In other words, if $\xi' = x_1 x_2 \cdots x_s x_{s+1}$,

$$x_1 \oplus x_2 \oplus \cdots \oplus x_s \oplus x_{s+1} = 0. \qquad (6.3)$$

This equality (and some easily verified properties of addition modulo 2) allow us to express x_{s+1} explicitly:

$$x_{s+1} = x_1 \oplus x_2 \oplus \cdots \oplus x_s. \qquad (6.4)$$

For example, if $\xi = 001011101$, then $\xi' = 0010111011$.

The words ξ' formed in this way define the set of intelligible words $N_2 \subseteq E(s + 1, \mathfrak{A}_2)$.

It is clear that the distance between two distinct intelligible words ξ' and ξ'' must be even, for if ξ' differed from ξ'' in an odd number of positions, then the sum of the units in one of the words ξ' or ξ'' would be odd, a situation made impossible by the construction of these words. And because the smallest even number not equal to zero is two, the minimum distance between distinct intelligible words is two; thus the subscript 2 is used.

Consequently, this code allows us to detect errors of one digit by counting the nonzero digits; if the evenness criterion (6.3) is not satisfied, then the word contains an error. This error detection process is widely known as the *evenness test* and is very frequently applied because of its simplicity. The redundancy of the error detection code is

$$\frac{(s + 1) - s}{s + 1} = \frac{1}{s + 1}.$$

We shall now consider a beautiful example of a code (due to Hemming) which is capable of correcting single-digit errors.

Let $\xi \in E(s, \mathfrak{A}_2)$ be a binary word of length s. We form the word $\xi' \in E(n, \mathfrak{A}_2)$ according to the following rule: Among the n positions in the word ξ', we choose the first, the second, the fourth, . . ., the (2^k)th positions for *controlling symbols* which are determined by the word ξ. Between these positions are the *significant positions*. In the example

$$\xi' = \mathbf{1}\mathbf{0}011\mathbf{1}00\mathbf{1}00110\mathbf{1}1111101100010010 \,,$$

we have indicated the mutual distribution of the controlling (distinguished by heavy type) and the significant positions for the case $s = 26$, $n = 31$, $k = 4$. In order to make s significant positions available, the number of controlling positions $(k + 1)$ plus the number s of significant positions must lie between the kth and $(k + 1)$st powers of 2; that is, it is necessary and sufficient that

$$2^k \leq s + k + 1 < 2^{k+1} . \tag{6.5}$$

The redundancy of the given code is $[(s + k + 1) - s]/s + k + 1 = (k + 1)/(s + k + 1)$.

The $(i + 1)$st controlling position (position 2^i) is filled according to the following rule: Each position of the word ξ' is defined by a number *l*, counting from the beginning of the word. We examine the binary representation of this number:

$$l = l_k 2^k + l_{k-1} 2^{k-1} + \cdots + l_1 2 + l_0$$

—the number of binary elements in the representation of the number *l* is defined so that, in accordance with (6.5), $l < 2^{k+1}$.

Let us now consider the set π_i of all those positions *l* for which $l_i = 1$. This set contains exactly one controlling position, the position with the number $l = 2^i$. We fill this position in such a way that the sum of all the ones in the positions of π_i will be even.

In table 6.1 we give an example of a word ξ', which can be read vertically in the second column. We have shown the binary number of the positions and marked with a star those positions belonging to the set π_i. Words ξ' constructed according to this rule shall be called *intelligible*.

We shall show that the distance between any two intelligible words ξ' and η' is not less than 3—that is, that the intelligible words form a set $N_3 \subseteq E(s + k + 1, \mathfrak{A}_2)$.

Case 1. Suppose words ξ and η from $E(s, \mathfrak{A}_2)$ differ in at least three positions. Clearly, then, the words ξ' and η' likewise differ in at least three positions and, consequently, $d(\xi', \eta') \geq 3$.

Case 2. Let the words ξ and η differ in two positions. Then the words ξ' and η' differ in two significant positions—say positions *l* and *l'*. Since $l \neq l'$, the binary representation of *l* differs from that of *l'* in at least one place. Let *i* be the place in which the two representations differ and, without loss of generality, say $l_i = 1$ and $l'_i = 0$. Then $l \in \pi_i$, but $l' \notin \pi_i$.

Because the words ξ' and η' differ in only two significant positions, and since $l' \notin \pi_i$, the sum of the significant digits in π_i for the word ξ' and the sum of digits in the corresponding set for η' must differ. As the sum of the digits in the set of positions π_i must be made even both for ξ' and for η', the words ξ' and η' must differ again in the controlling position of the set π_i (in position 2^i). Thus, ξ' and η' differ in at least three positions, and $d(\xi', \eta') \geq 3$.

Case 3. Let the words ξ and η differ in exactly one position. Then the words ξ' and η' differ in exactly one significant position with the number *l*. This number cannot be a power of two, since numbers of the form 2^i are used for the controlling positions. Therefore, the number *l* has at least two nonzero binary digits $l_i = 1$ and $l_j = 1$. Consequently, position *l* is in both π_i and π_j. Since the sum of the digits in these sets must be made even for both ξ' and η', ξ' and η' must differ in controlling positions 2^i and 2^j.

Table 6.1

Position No.	Contents of the Position	π_0	π_1	π_2	π_3	π_4
00001	1	*				
00010	0		*			
00011	0	*	*			
00100	1			*		
00101	1	*		*		
00110	1		*	*		
00111	0	*	*	*		
01000	0				*	
01001	1	*			*	
01010	0		*		*	
01011	0	*	*		*	
01100	1			*	*	
01101	1	*		*	*	
01110	0		*	*	*	
01111	1	*	*	*	*	
10000	1					*
10001	0	*				*
10010	1		*			*
10011	1	*	*			*
10100	1			*		*
10101	0	*		*		*
10110	1		*	*		*
10111	1	*	*	*		*
11000	0				*	*
11001	0	*			*	*
11010	0		*		*	*
11011	1	*	*		*	*
11100	0			*	*	*
11101	0	*		*	*	*
11110	1		*	*	*	*
11111	0	*	*	*	*	*

In each case, then, the words ξ' and η' differ in at least three positions, and $d(\xi', \eta') \geq 3$.

And so the set of intelligible messages is an N_3 set; consequently, one can in principle restore distorted messages even if the error occurs only in a single digit.

In the binary case, this process of restoration can be carried out with comparative ease, for to restore a word in the binary alphabet, it is sufficient to determine the number of the position in which the error has occurred and to change the entry in this position from 0 to 1 or vice versa.

In the code that we are examining, the number of the position with an incorrect entry can be ascertained by the following method: After the transmission of the message ξ', during which only one digit can be distorted (resulting in the message $\xi*$), we check whether the sum of the digits is even or odd for each set of positions π_i. In other words, we calculate the controlling quantities

$$\alpha_i = \xi_p^* \oplus \xi_q^* \oplus \xi_r^* \oplus \cdots ,$$

where α_i is equal to the sum modulo 2 of all the symbols in the positions of the set π_i of the received message $\xi*$.

If all of the $\alpha_i = 0$, then $\xi*$ is an intelligible message.[1] If, however, some $\alpha_i = 1$, then an error must exist in a position l belonging to the set π_i, that is, in a position where the ith binary digit is equal to 1. Conversely, for each $\alpha_j = 0$, no error occurs in any position belonging to the set π_j (since two errors that cancel each other out are extremely improbable).

Thus, the controlling quantities α_i are just the binary digits in the expansion of the number of the position in which the error has occurred; that is,

$$l = \alpha_k 2^k + \alpha_{k-1} 2^{k-1} + \cdots + \alpha_1 2 + \alpha_0 . \tag{6.6}$$

So the controlling quantities of the received message completely determine l and enable us to restore the correct message ξ'.

Let us take, for example, the word ξ' written in table 6 and distort it in the nineteenth position.

We obtain the word

$$\xi* = 1001110010011011010101100010010 ;$$

carrying out our test, we find

$$\alpha_4 = 1 , \quad \alpha_3 = 0 , \quad \alpha_2 = 0 , \quad \alpha_1 = 1 , \quad \alpha_0 = 1 ,$$

that is, $l = 10011$, the binary representation of the decimal number 19. Changing 0 to 1 in the nineteenth position of the word $\xi*$, we obtain ξ', the word that we started with.

1. That is, if an error exists, it is an error of at least three positions, a situation made impossible by the fact that the transmission time is so short that it is highly improbable for more than one error to occur.

A simpler code allowing us to correct single mistakes would result if the word ξ' were given by a triple repetition of the word $\xi \in E(s, \mathfrak{A}_2)$. Then if ξ and η differ in r positions, the corresponding words ξ' and η' would differ in $3r$ positions. Thus, $d(\xi', \eta') \geq 3$, provided that $\xi' \neq \eta'$. The transmitted word is checked in the following manner.

A triple of positions with the numbers $l, l + s, l + 2s$, where $1 \leq l \leq s$, is considered. If the symbols in these positions coincide, the corresponding symbol is considered to have been given correctly. Since the binary language contains only two letters, the symbols in two of these three positions must coincide; and so, if only two of these symbols coincide, their common meaning is considered to be the correct one and is entered in the third position.

Thus, this code is capable of correcting single errors in each triple of corresponding positions. The weakness of the code is its high redundancy, which is $(3s - s)/3s = 2s/3s = 2/3$. The redundancy of the former code is

$$\frac{k + 1}{s + k + 1} = \frac{[\log_2 (s + k + 1)] + 1}{s + k + 1},$$

where the square brackets denote the greatest integer function (since $2^k \leq s + k + 1 < 2^{k+1}$); setting the length of the word $(s + k + 1)$ equal to n, the redundancy becomes $\{[\log_2 n] + 1\}/n$, which, for large n (long words) is practically zero.

Codes that allow the correction of errors in the transmission and storage of information are very important in various automatic control devices. The last twenty years have seen the appearance of a great number of works concerning error-stabilizing codes that allow us to correct multiple as well as single errors.

7 Metrics and Norms in Multi-dimensional Spaces

In this chapter we examine an *n-dimensional vector space* $\mathbb{R}^n$ and various distance functions which determine metric spaces. The vector space $\mathbb{R}^n$ serves as a generalization of the concepts of line ($\mathbb{R}^1$), plane ($\mathbb{R}^2$), and three-space ($\mathbb{R}^3$) considered in elementary geometry. We can arrive at a reasonable definition of the *n*-dimensional space $\mathbb{R}^n$ (*n-space*) in the following manner.

We consider the plane with some system of Cartesian coordinates. Each point M on the plane is uniquely defined by a pair of coordinates (x, y), where $x \in \mathbb{R}$, $y \in \mathbb{R}$ (here $\mathbb{R}$ denotes the set of real numbers).

To each point M there corresponds in a one-to-one manner a *vector* joining the origin of the coordinate system to that point (see fig. 7.1). Thus, there exists a one-to-one correspondence between any two of the following objects:

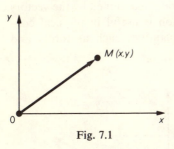

Fig. 7.1

The point $M \leftrightarrow$ the vector **OM** $\leftrightarrow$ the pair of coordinates (x, y). Consequently, we may think of the plane interchangeably as a set of points, a set of vectors, or a set of ordered pairs (x, y) of real numbers. Analogously, we can consider three-space as a set of ordered triples (x, y, z) of real numbers. Our desired definition (of *n*-space) suggests itself.

By a vector in *n*-space ($\mathbb{R}^n$) we mean an ordered *n*-tuple of real numbers

$$\xi = (x_1, x_2, \ldots, x_n).$$

The numbers $x_1, x_2, \ldots, x_n$ are called the *coordinates* of the vector ξ. The set of all such vectors is the *n-dimensional vector space* $\mathbb{R}^n$.

Clearly, the vector space $\mathbb{R}^3$ is ordinary three-space; the vector space $\mathbb{R}^2$ is the plane; and the space $\mathbb{R}^1$ is the straight line.

Three operations are defined on vectors in $\mathbb{R}^n$: addition, subtraction, and multiplication by a *scalar* (real number). These operations are defined as follows.

The *sum* of the vectors $\xi = (x_1, x_2, \ldots, x_n)$ and $\eta = (y_1, y_2, \ldots, y_n)$ is the vector

$$\zeta = \xi + \eta = (x_1 + y_1, x_2 + y_2, \ldots, x_n + y_n),$$

the coordinates of which are the sums of the corresponding coordinates of ξ and η.

Analogously, the *difference* of these same vectors ξ and η is the vector

$$\theta = \xi - \eta = (x_1 - y_1, x_2 - y_2, \ldots, x_n - y_n),$$

whose coordinates are the differences of the corresponding coordinates.

The *product* of the scalar a and the vector $\xi = (x_1, x_2, \ldots, x_n)$ is the vector

$$\varphi = a\xi = (ax_1, ax_2, \ldots, ax_n).$$

In other words, to multiply a vector by a scalar, we multiply each of the coordinates by the scalar. On the plane and in three-space these operations have natural geometrical interpretations. For the sake of clarity we shall examine two vectors, ξ and η, in the plane (fig. 7.2). From the diagram it is clear that the sum $\zeta = \xi + \eta$ is the vector formed by the diagonal of the parallelogram determined by the vectors ξ and η. This property of vector addition is useful in physical considerations involving sums of vector quantities such as forces and momenta.

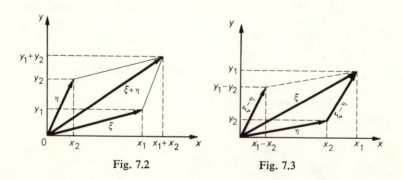

Fig. 7.2   Fig. 7.3

The difference of the vectors ξ and η (fig. 7.3) is the vector directed from the end of the vector η to the end of the vector ξ.

The product of the positive number a and the vector ξ is a vector having the same direction as ξ, but of length a times the length of ξ. (Clearly, when $a < 1$, the length of the vector $a\xi$ is less than that of the vector ξ.) To multiply the vector ξ by the negative number a, one must multiply it by $|a|$ and then take the vector with the same length but opposite direction. All of these cases are pictured in figure 7.4.

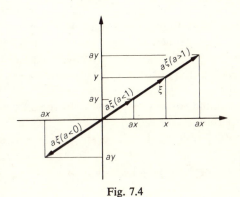

Fig. 7.4

One can easily verify that the operations on n-dimensional vectors defined above satisfy the following properties, which are analogous to the properties of the corresponding operations defined on the real numbers. Here ξ, $\eta \in \mathbb{R}^n$ and a, $b \in \mathbb{R}$. The symbol 0 is used to denote both the real number zero and the zero vector $(0, 0, \ldots, 0) \in \mathbb{R}^n$. When 0 is written as the sum or difference of vectors, the zero vector is denoted. All scalars are written to the left of vectors in a scalar multiplication.

1. $\xi + \eta = \eta + \xi$ (commutativity),
2. $\xi + (\eta + \zeta) = (\xi + \eta) + \zeta$ (additive associativity),
3. $\xi - \xi = 0$,
4. $0 + \xi = \xi$, where $0 \in \mathbb{R}^n$,
5. $a(\xi + \eta) = a\xi + a\eta$ (distributivity of scalar multiplication over vector addition),
6. $(a + b)\xi = a\xi + b\xi$ (distributivity of scalar multiplication over scalar addition),
7. $a(b\xi) = (ab)\xi$ (multiplicative associativity),
8. $0 \cdot \xi = 0$, where $0 \in \mathbb{R}$,
9. $a \cdot 0 = 0$, where $0 \in \mathbb{R}^n$,
10. $1 \cdot \xi = \xi$.

Clearly, two vectors, ξ and η, are equal if and only if $\xi - \eta = 0$.

Let us consider some other examples of multidimensional spaces that arise naturally in geometry.

Example 1. The set of all spheres in three-space. Each sphere is uniquely determined by an ordered 4-tuple (x, y, z, R), where (x, y, z) is the center and R the radius.

Example 2. The set of all triangles in three-space. Each triangle is uniquely determined by an ordered 9-tuple $(x_1, y_1, z_1, x_2, y_2, z_2, x_3, y_3, z_3)$, where the triple (x_i, y_i, z_i) gives the coordinates of the ith vertex of the triangle ($i = 1, 2,$ or 3). We suggest that the reader convince himself that in both of these examples multiplication of all of the coordinates (that is, of the corresponding four- or nine-dimensional vector) by the real number a is equivalent to performing a similarity transformation with the center of similarity at the origin. We further suggest that the reader make a more detailed study of various possible metrics in the spaces of spheres and triangles.

Let us now examine various metrics which we can define on $\mathbb{R}^n$ to form a metric space.

The metric d_2 (determining the metric space $l_2^{(n)}$) is defined by

$$d_2(\xi, \eta) = \sqrt{(x_1 - y_1)^2 + (x_2 - y_2)^2 + \cdots + (x_n - y_n)^2} \quad (7.1)$$

where

$$\xi = (x_1, x_2, \ldots, x_n) \quad \text{and} \quad \eta = (y_1, y_2, \cdots, y_n).$$

In three-space and in the plane, the metric d_2 is just the ordinary geometric distance function. Properties 1, 2, and 3 are obvious for this metric.

The space $l_1^{(n)}$ is defined by the metric d_1 where

$$d_1(\xi, \eta) = |x_1 - y_1| + |x_2 - y_2| + \cdots + |x_n - y_n|. \quad (7.2)$$

In the plane (the space $l_1^{(2)}$ defined by the set $\mathbb{R}^2$ and the metric d_1) this metric coincides with the metric d_1 defined in chapter 4. Again, properties 1, 2, and 3 are obvious.

The space $C^{(n)}$ results if we define a metric d_∞ according to the rule

$$d_\infty(\xi, \eta) = \max\left(|x_1 - y_1|, |x_2 - y_2|, \ldots, |x_n - y_n|\right), \quad (7.3)$$

that is, $d_\infty(\xi, \eta)$ is the maximal deviation of corresponding coordinates. Properties 1, 2, and 3 are obviously satisfied by this metric also. For the plane $\mathbb{R}^2$ it coincides with the metric d_∞ introduced in chapter 4.

Let us prove the triangle inequality for the space $l_1^{(n)}$.

Let

$$\xi = (x_1, x_2, \ldots, x_n); \qquad \eta = (y_1, y_2, \ldots, y_n); \qquad \zeta = (z_1, z_2, \ldots, z_n).$$

Then, obviously,

$$
\begin{aligned}
d_1(\xi, \eta) &= |x_1 - y_1| + |x_2 - y_2| + \cdots + |x_n - y_n| \\
&= |x_1 - z_1 + z_1 - y_1| + |x_2 - z_2 + z_2 - y_2| + \cdots \\
&\quad + |x_n - z_n + z_n - y_n| \\
&\leq |x_1 - z_1| + |z_1 - y_1| + |x_2 - z_2| + |z_2 - y_2| + \cdots \\
&\quad + |x_n - z_n| + |z_n - y_n| \\
&= d_1(\xi, \zeta) + d_1(\zeta, \eta),
\end{aligned}
$$

and the triangle inequality holds.

For the space $C^{(n)}$ with the metric d_∞, the triangle inequality is proved as follows. Let $|x_k - y_k|$ be the largest of the differences of corresponding coordinates; that is,

$$
\begin{aligned}
d_\infty(\xi, \eta) &= \max \left(|x_1 - y_1|, |x_2 - y_2|, \ldots, |x_n - y_n| \right) \\
&= |x_k - y_k| = |x_k - z_k + z_k - y_k| \\
&\leq |x_k - z_k| + |z_k - y_k|.
\end{aligned}
\tag{7.4}
$$

It is obvious that

$$
\left.
\begin{aligned}
|x_k - z_k| &\leq \max \left(|x_1 - z_1|, |x_2 - z_2|, \ldots, |x_n - z_n| \right) = d_\infty(\xi, \zeta), \\
|z_k - y_k| &\leq \max \left(|z_1 - y_1|, |z_2 - y_2|, \ldots, |z_n - y_n| \right) = d_\infty(\zeta, \eta).
\end{aligned}
\right\}
\tag{7.5}
$$

Combining (7.4) and (7.5), we get the desired relation

$$d_\infty(\xi, \eta) \leq d_\infty(\xi, \zeta) + d_\infty(\zeta, \eta).$$

A general class of metric spaces over $\mathbb{R}^n$ is obtained if we introduce a metric d_p defined by the formula

$$d_p(\xi, \eta) = \sqrt[p]{(x_1 - y_1)^p + (x_2 - y_2)^p + \cdots + (x_n - y_n)^p},$$

where $p \geq 1$; the space obtained in this way is called $l_p^{(n)}$.

Properties 1, 2, and 3 can be verified in this case exactly as before. The triangle inequality is derived from Minkowski's inequality (see the footnote on page 21):

$$
\begin{aligned}
\sqrt[p]{(a_1 - b_1)^p + (a_2 - b_2)^p + \cdots + (a_n - b_n)^p} \\
\leq \sqrt[p]{a_1^p + a_2^p + \cdots + a_n^p} + \sqrt[p]{b_1^p + b_2^p + \cdots + b_n^p}.
\end{aligned}
$$

It is not difficult to see that for $p = 1$ and $p = 2$ the spaces $l_1^{(n)}$ and $l_2^{(n)}$ defined above are obtained. As $p \to \infty$ the metric d_p approaches the metric d_∞; that is, $l_\infty^{(n)} = C^{(n)}$. The reader can easily verify this by generalizing the analogous argument in chapter 4.

A more difficult exercise is to show that the *sphere* of radius r in the space $l_1^{(3)}$ is an octahedron (fig. 7.5), whereas in the space $C^{(3)}$ it is a cube (fig. 7.6).

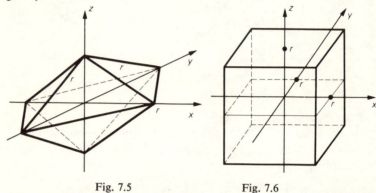

Fig. 7.5   Fig. 7.6

The spaces $l_p^{(n)}$ over $\mathbb{R}^n$, like the corresponding spaces l_p in the plane ($\mathbb{R}^2$), are called *Minkowski spaces*. These spaces can be generalized very naturally to an infinite-dimensional vector space whose elements are vectors with an infinite number of coordinates.

A more general class of metrics on $\mathbb{R}^n$ can be defined with the introduction of the concept of *convex body*.

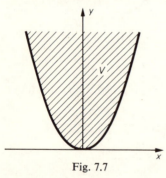

Fig. 7.7

Let us introduce several new definitions. We shall interpret a vector ξ in $\mathbb{R}^n$ as the point corresponding to the terminal point of that vector when the initial point is placed at the origin. We say that a subset V of $\mathbb{R}^n$ is *convex* if for each pair of vectors $\xi \in V$ and $\eta \in V$, all vectors of the form $a\xi + (1 - a)\eta$, where a is an arbitrary number between zero and one ($0 \le a \le 1$), are contained in V. Geometrically (in $\mathbb{R}^2$ and $\mathbb{R}^3$), this means that the entire line segment joining any two points of V is contained in V.

A subset $V \subset \mathbb{R}^n$ is *bounded* if there exists a positive number K such that for any $\xi = (x_1, x_2, \ldots, x_n) \in V$,

$$|x_1| < K; \ |x_2| < K; \ \cdots |x_n| < K.$$

Figure 7.7 pictures a convex but unbounded subset of $\mathbb{R}^2$; the subset pictured in figure 7.8 is both convex and bounded.

A vector ξ belonging to a subset V of $\mathbb{R}^n$ is said to be an *interior point* of V if for each vector $\eta \in \mathbb{R}^n$, there exists a positive number a such that $\xi + a\eta \in V$. In other words, if we move in an arbitrary direction from the end of the vector ξ, then we remain for some time in the set V.

If V is a flat surface in $\mathbb{R}^3$, then V has no interior points. In fact (see fig. 7.9), if $\xi \in V$ and the vector η is perpendicular to the plane in which V lies, then for any positive a the vector $\xi + a\eta$ lies outside this plane and, in particular, outside the set V.

A subset V of $\mathbb{R}^n$ is said to be *full-dimensional* if V has an interior point. A convex, bounded, full-dimensional subset V of $\mathbb{R}^n$ will be called a *convex body*.[1]

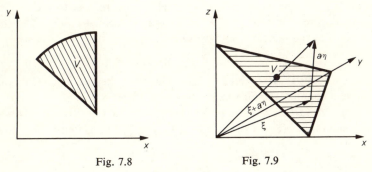

Fig. 7.8   Fig. 7.9

Let us now consider a convex body V *symmetric with respect to the origin* (that is, if $\xi \in V$, then $-\xi \in V$) in which the origin is an interior point.

This convex body can be used to define a metric d_V on $\mathbb{R}^n$. Suppose $\xi \in \mathbb{R}^n$, $\eta \in \mathbb{R}^n$; let $\zeta = \xi - \eta$. Since the origin $(0, 0, \ldots, 0)$ is an interior point of V, there exists a positive number a for which $a\zeta \in V$. It is easy to show that since V is bounded, there is an upper bound to $\{a \mid a\zeta \in V\}$. In other words, there is a lower bound to $\{1/a \mid a\zeta \in V\}$.

At this point, we introduce the concept of *greatest lower bound*. If $A \subset \mathbb{R}$ is bounded below, then the greatest lower bound of A (written $\inf_{a \in A} a$) is the unique number b for which the following conditions are satisfied:[2]

1. We leave it to the reader to show that a convex body must have infinitely many interior points.

2. The question of the uniqueness and existence of the greatest lower bound is involved and need not concern us here. For a full discussion, see Walter Rudin, *Principles of Mathematical Analysis* (New York: McGraw-Hill, 1964).

1. If $a \in A$, then $b \le a$;
2. If c is a lower bound for A, then $c \le b$.

We are now in a position to define the distance $d_V(\xi, \eta)$ as the greatest lower bound of $\{1/a \mid a > 0, a\zeta \in V\}$ (recall that $\zeta = \xi - \eta$), that is,

$$d_V(\xi, \eta) = \inf_{a\zeta \in V} \frac{1}{a}. \qquad (7.6)$$

For $\zeta = 0$, that is, when the vectors ξ and η are equal, any a is admitted (as we always have $a\zeta = 0 \in V$), and the greatest lower bound in (7.6) is zero. When ξ and η do not coincide, the vector $\zeta \ne 0$ and the permissible values of a are bounded from above by some positive number A. Therefore, the values of $1/a$ are bounded below by $1/A$, so $d_V(\xi, \eta) = \inf_{a\zeta \in V} 1/a > 0$. In other words, if $\xi \ne \eta$, $d_V(\xi, \eta)$ is strictly greater than zero.

Because the convex body V is symmetric with respect to the origin, $a\zeta \in V$ if and only if $-a\zeta = a(-\zeta) \in V$. Since $-\zeta = \eta - \xi$, the symmetry of the metric is shown; that is,

$$d_V(\xi, \eta) = d_V(\eta, \xi).$$

Thus, properties 1, 2, and 3 (page 12) hold for the metric d_V. Only the triangle inequality remains to be shown.

Let ξ, η, and ζ be vectors in $\mathbb{R}^n$. We choose two positive numbers a and b such that $a(\xi - \zeta) \in V$ and $b(\zeta - \eta) \in V$. Let us denote by α the quantity

$$\alpha = \frac{\dfrac{1}{a}}{\dfrac{1}{a} + \dfrac{1}{b}} = \frac{b}{a + b}.$$

Clearly, $0 < \alpha < 1$ and $1 - \alpha = a/(a + b)$. Because V is convex, the vector

$$\theta = \alpha[a(\xi - \zeta)] + (1 - \alpha)[b(\zeta - \eta)] \qquad (7.7)$$

is also contained in V. Transforming expression (7.7) by substituting the values of α and $1 - \alpha$ and using the properties of operations on vectors, we get

$$\theta = \frac{b}{a + b} \cdot a(\xi - \zeta) + \frac{a}{a + b} \cdot b(\zeta - \eta)$$

$$= \frac{ab}{a + b} [\xi - \zeta + \zeta - \eta] = \frac{ab}{a + b} (\xi - \eta) = c(\xi - \eta), \quad (7.8)$$

where $c = ab/(a + b)$.

Since the vector $\theta = c(\xi - \eta) \in V$, the number

$$\frac{1}{c} = \frac{a+b}{ab} = \frac{1}{b} + \frac{1}{a}$$

must be at least as great as the distance $d_V(\xi, \eta)$ (by the definition of $d_V(\xi, \eta)$ as $\inf\limits_{a(\xi - \eta) \in V} 1/a$):

$$d_V(\xi, \eta) \leq \frac{1}{a} + \frac{1}{b} \, . \tag{7.9}$$

However, by the definition of the numbers a and b, the quantities $1/a$ and $1/b$ can be made arbitrarily close to the distances $d_V(\xi, \zeta)$ and $d_V(\zeta, \eta)$, respectively, since $d_V(\xi, \zeta)$ is the greatest lower bound of the $1/a$ and $d_V(\zeta, \eta)$ is the greatest lower bound of the $1/b$. And since the inequality is preserved in the limiting process, (7.9) yields the desired inequality

$$d_V(\xi, \eta) \leq d_V(\xi, \zeta) + d_V(\zeta, \eta) \, . \tag{7.10}$$

It is possible to develop the definition of the metric d_V in several other ways.

The *norm* of the vector ξ is the quantity

$$\|\xi\|_V = d_V(\xi, 0) = \inf\limits_{a\xi \in V} \frac{1}{a} \, . \tag{7.11}$$

It is clear that the distance $d_V(\xi, \eta)$ defined above is just the norm of the difference of the vectors ξ and η, that is,

$$d_V(\xi, \eta) = \|\xi - \eta\|_V \, . \tag{7.12}$$

It can be shown (and we leave the proof to the reader) that this norm satisfies the following properties for $\xi \in \mathbb{R}^n$, $\eta \in \mathbb{R}^n$, $a \in \mathbb{R}$:

1. $\|\xi\|_V \geq 0$;
2. $\|\xi\|_V = 0$ if and only if $\xi = 0$;
3. $\|a\xi\|_V = |a| \, \|\xi\|_V$;
4. $\|\xi + \eta\|_V \leq \|\xi\|_V + \|\eta\|_V$.

It is possible to arrive at the concept of norm by a more abstract route. We define a norm on the n-dimensional vector space $\mathbb{R}^n$ as a function $U: \mathbb{R}^n \to \mathbb{R}$ (where $\|\xi\|$ denotes $U(\xi)$) defined for each $\xi \in \mathbb{R}^n$ in such a way that the following properties are satisfied (here $\xi \in \mathbb{R}^n$, $\eta \in \mathbb{R}^n$, $a \in \mathbb{R}$):

1. $\|\xi\| \geq 0$;
2. $\|\xi\| = 0$ if and only if $\xi = 0$;
3. $\|a\xi\| = |a|\,\|\xi\|$;
4. $\|\xi + \eta\| \leq \|\xi\| + \|\eta\|$.

The vector space $\mathbb{R}^n$ with a norm defined on it is called an *n-dimensional Minkowski space*.[3] It can be shown that every norm can be defined by some symmetric convex body V. To verify this assertion, we consider the set V consisting of all vectors ξ for which $\|\xi\| \leq 1$. We shall first show that V is a symmetric convex body in $\mathbb{R}^n$.

To show that V is convex, let $\xi \in V$ and $\eta \in V$, and let $0 \leq a \leq 1$. Then, by properties 3 and 4 of norms,

$$\|a\xi + (1 - a)\eta\| \leq \|a\xi\| + \|(1 - a)\eta\| = a\|\xi\| + (1 - a)\|\eta\|\,.$$

Since $\|\xi\| \leq 1$ and $\|\eta\| \leq 1$, we have

$$\|a\xi + (1 - a)\eta\| \leq a + (1 - a) = 1\,,$$

that is, the vector $a\xi + (1 - a)\eta$ also belongs to the set V; thus, V is convex.

Second, we must show that the origin is an interior point of the set V. If ξ is an arbitrary nonzero vector, we obtain upon setting $a = 1/\|\xi\|$ that

$$\|a\xi\| = a\|\xi\| = \frac{1}{\|\xi\|}\,\|\xi\| = 1\,;$$

that is, $a\xi \in V$. If $\xi = 0$, then for any positive real number a, $a\xi \in V$, since by property 2, $\|a\xi\| = \|0\| = 0 \leq 1$.

The symmetry of the set V follows from property 3; if $\xi \in V$, $\|-\xi\| = \|(-1)\xi\| = |-1|\,\|\xi\| = \|\xi\| \leq 1$. So if $\xi \in V$, then $-\xi \in V$.

The proof of the boundedness of the set V is somewhat more cumbersome, and so we shall omit it.

Since the set V is a convex body, it defines a norm n, denoted by $n(\xi) = \|\xi\|_V$. It still remains to be shown that this norm coincides with the one chosen at the beginning of the proof. Let a be an arbitrary positive number for which $a\xi \in V$. This means that $\|a\xi\| \leq 1$, which

3. In honor of the great mathematician H. Minkowski, one of the creators of the theory of relativity.

implies that $a\|\xi\| \leq 1$, or $1/a \geq \|\xi\|$. So if $a\xi \in V$, $1/a \geq \|\xi\|$, and the least such $1/a$ is obtained by setting $1/a$ strictly equal to $\|\xi\|$; that is, by setting $a = 1/\|\xi\|$. In other words, the greatest lower bound of the $1/a$ is $\|\xi\|$, or

$$\inf_{a\xi \in V} \frac{1}{a} = \|\xi\|. \qquad (7.13)$$

Comparing equations (7.13) and (7.11), we see that

$$\|\xi\| = \|\xi\|_V, \qquad (7.14)$$

the desired result.

In $\mathbb{R}^3$, the norm defined by a given convex body V has a simple physical interpretation.

Let us suppose that we have some anisotropic device which propagates sound waves at various speeds in different directions, and consider the case in which the speed of sound in opposite directions is the same.

We now choose a unit of speed (such as miles per hour) and construct in each direction from the origin a vector whose length is equal to the speed of sound in that direction. We make the further assumption that the terminal points of these vectors bound a convex body V. Clearly, V is bounded, symmetric with respect to the origin, and contains at least one interior point, the origin. So there is a norm $\|\xi\|_V$ and a metric d_V defined by

$$d_V(\xi, \eta) = \|\xi - \eta\|_V.$$

We leave it to the reader to verify that the distance $d_V(\xi, \eta)$ is numerically equal to the time required for a sound wave to travel from the end of the vector ξ to the end of the vector η along the straight line connecting them.

In addition to the finite-dimensional Minkowski spaces, one can consider their infinite-dimensional analogs—the so-called Banach spaces.[4] In general, a Banach space is a vector space on which a norm can be defined (that is, a space satisfying all of the properties listed on page 49 along with a norm possessing all of the properties listed on page 56).

4. In honor of the Polish mathematician S. Banach (1892–1945), one of the founders of functional analysis—an important branch of modern mathematics.

We can construct an example of an infinite-dimensional Banach space in the following way: Let $C([0, 1])$ be the set of all continuous functions on the closed interval $[0, 1] = \{t \mid 0 \le t \le 1\}$. The sum of two such vectors (in this case, functions) is defined by its operation on a number t; that is, $f + g$ is defined by $(f + g)(t) = f(t) + g(t)$ for f and g elements of $C([0, 1])$. Similarly, the scalar product af (for $a \in \mathbb{R}$) is defined by $(af)(t) = a[f(t)]$. The zero vector is the function 0 defined by $0(t) = 0$, all $t \in [0, 1]$. As the norm of a function we take the maximum of the absolute values of elements of the range, that is,

$$\|f\| = \max_{0 \le t \le 1} |f(t)| \, .$$

(It can be shown that because of the choice of the domain $[0, 1]$, such a maximum necessarily exists.)

It is easy to show that all of the previously listed vector space and norm properties are satisfied by the space $C([0, 1])$ with the norm defined above.

Banach spaces in which the vectors are functions play an important role in modern mathematics.

The claim that metric spaces whose points are functions are infinite-dimensional can in some sense be justified by the following reasoning.

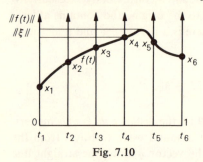

Fig. 7.10

We partition the closed interval $[0, 1]$ by drawing vertical lines through n of its points (fig. 7.10). Now we take the vector $\xi = (x_1, x_2, \ldots, x_n) \in \mathbb{R}^n$ and represent its coordinates on these vertical lines. The points in the plane determined in this way form the graph of some function defined on the n chosen points. Clearly, as $n \to \infty$, this set of points "converges" to the graph of continuous function if we have chosen points in $\mathbb{R}^n$ whose coordinates on adjacent vertical lines become arbitrarily close as $n \to \infty$. If we define a norm on $\mathbb{R}^n$ by

$$\|\xi\| = \max \, (|x_1|, |x_2|, \ldots, |x_n|)$$

(where $\xi = (x_1, x_2, \ldots, x_n) \in \mathbb{R}^n$), then this norm "in the limit" (as $n \to \infty$) becomes the norm defined by

$$\|f\| = \max_{0 \le t \le 1} |f(t)| ,$$

where $f \in C([0, 1])$.

The point is that n, the dimension of the normed space in question, increases without bound, indicating that the "limiting" space $C([0, 1])$ is infinite-dimensional.

8

The Smoothing of Errors in Experimental Measurements

In the measurement of physical quantities, experimental results often appear as a sequence $(x_1, x_2, \ldots, x_n)$ of observed values.

The quantity itself can be constant or variable. In the latter case, the values $x_1, x_2, \ldots, x_n$ should vary according to some law; in the former case, they should be nearly equal. But in any case, the measured quantities $x_1, x_2, \ldots, x_n$ are subject to error. In other words, there are inherent experimental imperfections that hinder the reception of information from nature.

The mathematical problem concerned with the treatment of measurements is that of the establishment (so far as possible) of the correct information. The solution lies in the application of concepts developed previously for the automatic correction of errors in discrete messages.

If the measured quantities can take on arbitrary real values, we can consider the n-dimensional vector space $\mathbb{R}^n$ as our space of information. The distance $d(\xi, \eta)$ between points of this space can be defined to fit the experiment being carried out. But most frequently, a metric d of the form

$$d(\xi, \eta) = \sqrt{(x_1 - y_1)^2 + \cdots + (x_n - y_n)^2} \qquad (8.1)$$

is used, for which the space of information is $l_2^{(n)}$.

Let $N \subset \mathbb{R}^n$ be a subset of this space of information.

As a "correct" message, we take the vector $\eta \in N$ "closest" to the message ξ that is received, that is, a vector η such that

$$d(\xi, \eta) = \min_{\eta' \in N} d(\xi, \eta') \qquad (8.2)$$

(if such a vector exists). It can be shown that in the interesting cases (for example, when the set N consists of all vectors which require the x_i to lie along some curve when plotted against time coordinates t_i) such a minimum exists, and thus so does the vector η. For the metric defined by (8.1), this principle is commonly known as the *method of least squares*, a method introduced by the great German mathematician Karl Friedrich Gauss.

Let us examine a concrete example of a subset of theoretically possible messages. We suppose that the measured quantity changes linearly with respect to time, that is, if y is the measured quantity,

$$y = kt + b,$$

where k and b are some constants and t is the time.[1]

This means that each vector $\eta \in N$ has the form $\eta = (y_1, y_2, \ldots, y_n)$, where

$$y_1 = kt_1 + b,$$
$$y_2 = kt_2 + b,$$
$$\cdot \quad \cdot \quad \cdot \quad \cdot \quad \cdot \quad \cdot$$
$$y_n = kt_n + b.$$

Let the vector actually obtained by measuring this quantity be equal to $\xi = (x_1, x_2, \ldots, x_n)$. The fundamental condition (8.2) is now written as follows:

$$F(k, b) = (kt_1 + b - x_1)^2 + (kt_2 + b - x_2)^2 + \cdots + (kt_n + b - x_n)^2$$
$$= \min.$$

In the expression $F(k, b)$, the unknowns are the parameters k and b defining the unknown theoretically possible messages; the quantities $t_1, t_2, \ldots, t_n$ and $x_1, x_2, \ldots, x_n$ are experimentally known.

To find the minimum value of the quantity $F(k, b)$, we use a criterion from differential calculus:

$$\frac{\partial F}{\partial k} = 0; \qquad \frac{\partial F}{\partial b} = 0, \tag{8.3}$$

1. For the sake of simplicity, we shall assume that the error involved in defining moments of time t_i is negligible.

which, in the given case of a positive quadratic function $F(k, b)$, is necessary and sufficient for a minimum.

Let us calculate the partial derivatives:

$$\left.\begin{array}{l} \dfrac{\partial F}{\partial k} = 2t_1(kt_1 + b - x_1) + \cdots + 2t_n(kt_n + b - x_n), \\[2mm] \dfrac{\partial F}{\partial b} = 2(kt_1 + b - x_1) + \cdots + 2(kt_n + b - x_n). \end{array}\right\} \tag{8.4}$$

For convenience, we denote

$$[t^2] = t_1{}^2 + t_2{}^2 + \cdots + t_n{}^2,$$

$$[t] = t_1 + t_2 + \cdots + t_n,$$

$$[tx] = t_1x_1 + t_2x_2 + \cdots + t_nx_n,$$

$$[x] = x_1 + x_2 + \cdots + x_n,$$

$$[1] = 1 + 1 + \cdots + 1 = n.$$

The expression (8.4) can then be written in the form

$$\left.\begin{array}{l} \dfrac{\partial F}{\partial k} = 2[t^2]k + 2[t]b - 2[tx], \\[2mm] \dfrac{\partial F}{\partial b} = 2[t]k + 2[1]b - 2[x]. \end{array}\right\} \tag{8.5}$$

Setting these expressions equal to zero in accordance with (8.3), dividing by two, and transferring the free terms to the right side, we get the fundamental equation of the method of least squares in symbolic form:

$$\left.\begin{array}{l} [t^2]k + [t]b = [tx], \\[2mm] [t]k + [1]b = [x]. \end{array}\right\} \tag{8.6}$$

Figure 8.1 pictures measured values $x_1, x_2, x_3, x_4, x_5, x_6, x_7, x_8$, and the

Table 8.1

	x_i	t_i	$t_i x_i$	t_i^2
1	0.20	0.30	0.06	0.09
2	0.43	0.91	0.39	0.83
3	0.35	1.50	0.53	2.25
4	0.52	2.00	1.04	4.00
5	0.81	2.20	1.78	4.84
6	0.68	2.62	1.79	6.86
7	1.15	3.00	3.45	9.00
8	0.85	3.30	2.81	10.89
Σ	$[x] = 4.79$	$[t] = 15.83$	$[tx] = 11.85$	$[t^2] = 38.76$

straight line $y = kt + b$ defined according to the method of least squares. The figure makes it clear why we speak of the "smoothing" of errors.

Table 8.1 shows the order in which the calculation is carried out. The system (8.6) in this case has the form

$$38.76k + 15.83b = 11.85 ,$$

$$15.83k + 8b = 4.79 .$$

The solution is $k = 0.319$, $b = -0.032$. The unknown "message" is $y = 0.319t - 0.032$.

Analogously, if the set of theoretically possible messages N consists of all parabolic functions of the form $y = at^2 + bt + c$, then the fundamental condition (7.11) can be written in the form

$$F(a, b, c) = (at_1^2 + bt_1 + c - x_1)^2 + \cdots + (at_n^2 + bt_n + c - x_n)^2$$
$$= \min .$$

The minimizing of the functions $F(a, b, c)$ reduces to the solution of a system of three linear equations in three unknown parameters a, b, and c.

The method of least squares can also be easily carried out in the case of a metric d of the form

$$d(\xi, \eta) = \alpha_1(x_1 - y_1)^2 + \alpha_2(x_2 - y_2)^2 + \cdots + \alpha_n(x_n - y_n)^2 , \quad (8.7)$$

where $\xi = (x_1, x_2, \ldots, x_n)$ and $\eta = (y_1, y_2, \ldots, y_n)$ are elements of $\mathbb{R}^n$, and $\alpha_1, \alpha_2, \ldots, \alpha_n$ are positive real numbers (weights). Unequal weights must be employed if it is known that separate measurements in the experiment are not equally exact. In this case, it is necessary to assign smaller weights to less accurate measurements.

The fundamental principle (8.2) for the smoothing of errors can also be applied to the metrics of the spaces $C^{(n)}$ and $l_1^{(n)}$. However, in these cases, methods of determining the set of theoretically possible messages are more complicated.

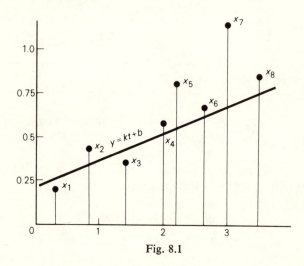

Fig. 8.1

9

A More
General Definition
of Distance

As we have already stated, various generalizations of the notion of distance are possible. One of the most radical is used in the theory of relativity, where we consider the space-time universe consisting of points of the form (x, y, z, t), where x, y, and z are spatial coordinates and t is the time coordinate. The distance (space-time interval) between two such points is defined by the formula

$$d(\xi, \eta) = \sqrt{c^2(t - t_1)^2 - (x - x_1)^2 - (y - y_1)^2 - (z - z_1)^2}, \quad (9.1)$$

where c is the speed of light. It is clear that the metric d can assume imaginary as well as real values.

It is also possible to generalize the concept of distance by assuming that a function d, satisfying axioms 1, 2, 3, and 4 (chap. 3, page 12), can have infinite value.[1] In this case, however, the space could be partitioned into disjoint subsets, each of which would be a metric space in the usual sense. Consequently, such a generalization is not very interesting. The proof of this fact can be sketched as follows.

Examining such a "generalized" space E, let us say two elements M and N of E are "equivalent" if the distance $d(M,N)$ is finite. Then, clearly, each point M is equivalent to itself, and if M is equivalent to N, N is equivalent to M $(d(N,M) = d(M,N))$. If M is equivalent to L and L is equivalent to N, then since

$$d(M,N) \le d(M,L) + d(L,N),$$

$d(M,N)$ is finite and M is equivalent to N. Thus the relation of "equivalence" partitions the space E into "equivalence classes," each of which is an ordinary metric space with a finite distance function.

1. More precisely, the value $+\infty$.

What this example points out is that it is no simple matter to come up with a meaningful generalization of an abstract mathematical concept. In every case, such a generalization must come from a deep study of the mathematical objects involved and not simply from a formal manipulation of axioms. The abortive attempt described above notwithstanding, there do exist a number of meaningful generalizations of the concept of metric space, one of which we shall study further. By giving up the axiom of symmetry (axiom 1), we obtain a class of spaces which is connected with some interesting mathematical objects.

We shall define a *generalized metric space* to be a set E and a function $d: E \times E \to \mathbb{R}$ (meaning that d has as its domain the Cartesian product of E with itself and as its range the real numbers) with the following properties (here ξ, η, and ζ are elements of E):

1. $d(\xi, \eta) \geq 0$.

2. The double equality $d(\xi, \eta) = d(\eta, \xi) = 0$ is satisfied if and only if $\xi = \eta$.

3. $d(\xi, \eta) \leq d(\xi, \zeta) + d(\zeta, \eta)$.

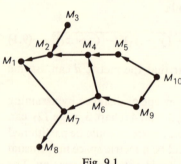

Fig. 9.1

Clearly, any ordinary distance function satisfies these conditions. However, a function nonsymmetric with respect to its arguments can also satisfy axioms 1, 2, and 3. In fact, we introduced such a nonsymmetric distance function at the end of chapter 4 in connection with the definition of distance as the minimal time required for travel from one point to another. Since a journey in the opposite direction may require more time, this metric is, in general, not symmetric, but the triangle inequality (and axioms 1 and 2) are easily verified.

Another nonsymmetric distance function is definable on the space consisting of the ten vertices of the diagram in figure 9.1.

The distance $d(M_i, M_j)$ between the points M_i and M_j is defined as the minimal number of line segments passing against the arrows in a path joining M_i and M_j.

For example,

$$d(M_1, M_{10}) = 4 ; \qquad d(M_{10}, M_1) = 0 ;$$

$$d(M_3, M_9) = 3 ; \qquad d(M_9, M_3) = 1 ; \quad \text{and so on}.$$

Clearly, $d(M_i, M_j) \geq 0$. The condition $d(M_i, M_j) = d(M_j, M_i) = 0$ (both distances being zero) means that it is possible to join the point M_i to the point M_j and to join M_j to M_i by means of line segments directed with the arrows; that is, M_i and M_j are vertices of a closed path on which all arrows go the same way. As there are no such loops in figure 9.1, the equality $d(M_i, M_j) = d(M_j, M_i) = 0$ implies that the points M_i and M_j coincide. Thus, conditions 1 and 2 hold for this nonsymmetric distance function.

The triangle inequality can be verified by the following argument. We examine a path with the minimal number of segments directed against the arrows joining M_i to M_k and an analogous path from M_k to M_j. Joining these paths, we obtain a path from M_i to M_j with the number of line segments directed against the arrows equal to $d(M_i, M_k) + d(M_k, M_j)$. Since in the "shortest" path from M_i to M_j, the number of such segments is at least as small,

$$d(M_i, M_j) \leq d(M_i, M_k) + d(M_k, M_j). \tag{9.2}$$

In this example it is possible to define a new metric d^* by the rule

$$d^*(M_i, M_j) = d(M_i, M_j) + d(M_j, M_i). \tag{9.3}$$

Clearly $d^*(M_i, M_j)$ possesses the properties of an ordinary metric. The analogous proposition is true in an arbitrary generalized metric space.

THEOREM *If (S, d) is a generalized metric space and $d^*: S \times S \to \mathbb{R}$ is defined by*

$$d^*(\xi, \eta) = d(\xi, \eta) + d(\eta, \xi), \tag{9.4}$$

then (S, d^*) is a metric space in the ordinary sense.

Proof. The symmetry of the metric d^* follows because the right side of (9.4) does not change upon interchanging ξ and η. The equality $d^*(\xi, \eta) = 0$ is equivalent to the double equality $d(\xi, \eta) = d(\eta, \xi) = 0$ (since d takes on no negative values) and is therefore equivalent to the statement $\xi = \eta$. Finally, since

$$d(\xi, \eta) \leq d(\xi, \zeta) + d(\zeta, \eta)$$

and

$$d(\eta, \xi) \leq d(\eta, \zeta) + d(\zeta, \xi),$$

we get

$$d(\xi, \eta) + d(\eta, \xi) \leq d(\xi, \zeta) + d(\zeta, \xi) + d(\zeta, \eta) + d(\eta, \zeta),$$

or

$$d^*(\xi, \eta) \leq d^*(\xi, \zeta) + d^*(\zeta, \eta),$$

and so the triangle inequality holds for the metric d^*.

Another interesting example of a generalized metric space can be obtained using the important concept of a *partially ordered set*.

A set S is said to be partially ordered if for some ordered pairs of points $(M, N) \in S \times S$, the relation $M \mathbin{(\!(} N$ (read M precedes N) is defined and satisfies the following axioms:

1. If $M \mathbin{(\!(} N$ and $N \mathbin{(\!(} M$, then $M = N$ (antisymmetry).
2. If $M \mathbin{(\!(} N$ and $N \mathbin{(\!(} L$, then $M \mathbin{(\!(} L$ (transitivity).

An example of a partially ordered set is the set of vertices in figure 9.1. We set $M_j \mathbin{(\!(} M_i$ if there is a path joining M_i to M_j which moves only in the direction of the arrows. For example, $M_8 \mathbin{(\!(} M_{10}, M_1 \mathbin{(\!(} M_3, M_2 \mathbin{(\!(} M_6, M_1 \mathbin{(\!(} M_1$.

A second example is obtained by considering $\mathbin{(\!(}$ to denote the relation " $<$ " on the real line; that is, $x \mathbin{(\!(} y$ if and only if $x < y$. In this case, it is clear that for each pair of distinct points x and y either $x \mathbin{(\!(} y$ or $y \mathbin{(\!(} x$ is valid. A set with such an ordering (in which for each pair of distinct points x and y, either $x \mathbin{(\!(} y$ or $y \mathbin{(\!(} x$) is said to be *linearly ordered* by $\mathbin{(\!(}$.

In any partially ordered set it is possible to introduce the notion of an *immediate predecessor*.

A point M is said to immediately precede a point N (and we write $M \bigcirc N$) if $M \mathbin{(\!(} N$ and there is no third point L different from M and N lying "between" M and N; that is, such that $M \mathbin{(\!(} L \mathbin{(\!(} N$.

For example, in figure 9.1, $M_2 \bigcirc M_3, M_2 \bigcirc M_4, M_1 \bigcirc M_7$, and so on.

No real number has an immediate predecessor, for if $x \mathbin{(\!(} y$, then $x \mathbin{(\!(} (x + y)/2 \mathbin{(\!(} y$, since $x < (x + y)/2 < y$.

We now consider a finite partially ordered set E and suppose that that set has the property of *connectedness*; that is, for each pair of points M and N in E there exists a sequence of points $M = L_1, L_2, \ldots, L_k = N$ such that for each i with $1 \leq i \leq k - 1$, either $L_i \mathbin{(\!(} L_{i+1}$ or $L_{i+1} \mathbin{(\!(} L_i$. For the points M_3 and M_8 in figure 9.1, for example, we can construct such a sequence as follows:

$$L_1 = M_3; \quad L_2 = M_1; \quad L_3 = M_7; \quad L_4 = M_8,$$

since

$$M_1 \,(\!(\, M_3\, ; \qquad M_7 \,(\!(\, M_1\, ; \qquad M_8 \,(\!(\, M_7\, .$$

We leave it to the reader to check that the set of vertices in figure 9.1 is connected.

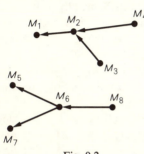

The set of points in figure 9.2 is not connected, since for the points M_8 and M_3 such a connecting sequence does not exist. However, the subsets $E_1 = \{M_1, M_2, M_3, M_4\}$ and $E_2 = \{M_5, M_6, M_7, M_8\}$ are connected. One can easily verify that any finite partially ordered set can be partitioned into disjoint connected subsets.

Fig. 9.2

We are now in a position to introduce into an arbitrary partially ordered set E a metric d defined according to the following rule. We first define a *path* from a point M to a point N to be a chain of points $M = L_1, L_2, \ldots, L_k = N$ such that for each i with $1 \leq i \leq k - 1$, either $L_i \bigcirc L_{i+1}$ or $L_{i+1} \bigcirc L_i$. We then define the distance $d(M, N)$ to be the length of the shortest path from M to N—the length of a path being defined as the number of integers i such that $1 \leq i \leq k - 1$ and $L_{i+1} \bigcirc L_i$ (that is, the number of steps "against the arrows").

The nonnegativity of the distance $d(M, N)$ follows from the definition. For the proof of the second axiom of distance we note that for distinct points M and N, the condition $d(M, N) = 0$ implies that there exists a chain of points $M = L_1, L_2, \ldots, L_k = N$ such that $L_i \bigcirc L_{i+1}$ and, in particular, that $L_i \,(\!(\, L_{i+1}$ for each i. But then, by the second axiom (transitivity) for partially ordered sets, we get that $M \,(\!(\, N$. Analogously, from the condition $d(N, M) = 0$, it follows that $N \,(\!(\, M$. Thus, if $d(M, N) = d(N, M) = 0$ is satisfied, $M \,(\!(\, N$ and $N \,(\!(\, M$; so, by the first axiom (antisymmetry) for partially ordered sets, $M = N$. Conversely, if $M = N$, then the length $d(M, N)$ of the shortest path from M to N and the length $d(N, M)$ of the shortest path from N to M are equal to zero. So the metric d satisfies the second condition for a generalized metric.

For the proof of the third axiom (the triangle inequality) we employ a familiar method. Taking a shortest path $M = L_1, L_2, \ldots, L_k = Q$ from M to Q and a shortest path $Q = L_k, L_{k+1}, \ldots, L_{k+p} = N$ from Q to N, we form a path $M = L_1, L_2, \ldots, L_k = Q = L_k, L_{k+1}, \ldots, L_{k+p}$

$= N$ from M to N. The total number of pairs of adjacent points in this chain for which $L_{i+1} \bigcirc L_i$ is equal to the sum of the distances $d(M, Q)$ and $d(Q, N)$. Clearly, the number of such pairs in the shortest path from M to N can only be smaller:

$$d(M, N) \leq d(M, Q) + d(Q, N). \tag{9.5}$$

And we have shown that for any finite connected partially ordered set E we can define a metric d so that (E, d) forms a metric space.

As an exercise, we suggest that the reader prove that for every pair of points M and N where $M \llparenthesis N$, the distance $d(M, N) = 0$.

In a sense, the converse assertion is also true. In any generalized metric space (E, d) it is possible to introduce a partial ordering $\llparenthesis$ defined by $M \llparenthesis N$ if $d(M, N) = 0$.

To prove this, we must show that both axioms for a partial ordering are satisfied. If $M \llparenthesis N$ and $N \llparenthesis M$, then $d(M, N) = d(N, M) = 0$, and by the second condition for a generalized metric, $M = N$. The first axiom for partially ordered sets is thus proved.

Now let $M \llparenthesis L$ and $L \llparenthesis N$. Then $d(M, L) = 0$ and $d(L, N) = 0$. By the triangle inequality

$$d(M, N) \leq d(M, L) + d(L, N) = 0;$$

but by nonnegativity, $0 \leq d(M, N)$, and so $0 \leq d(M, N) \leq 0$; that is, $d(M, N) = 0$ and $M \llparenthesis N$.

Thus, we have shown that if $M \llparenthesis L$ and $L \llparenthesis N$, then $M \llparenthesis N$.

The partially ordered set which we obtain in this way need not be connected.

Examining, for example, the partially ordered set E of the points in figure 9.2, one can define between pairs of points from the subset $E_1 = \{M_1, M_2, M_3, M_4\}$ a distance by means of the shortest path. The same can be done for the subset $E_2 = \{M_5, M_6, M_7, M_8\}$. We further define the distance between a point $M_i \in E_1$ and a point $M_j \in E_2$ by

$$d(M_i, M_j) = d(M_j, M_i) = 100. \tag{9.6}$$

It is easy to verify that we get a generalized metric space in which the equality $d(M, N) = 0$ is equivalent to the relation $M \llparenthesis N$ in the partially ordered set E. As we have already remarked, however, E is not a connected set.

It is possible, however, to introduce the notion of a *connected generalized metric space*. The space (E, d) is said to be connected if for any pair of (not necessarily distinct) points M and N of E there exists a chain of points $M = L_1, L_2, \ldots, L_k = N$ such that for each adjacent pair of points L_i and L_{i+1} either $d(L_i, L_{i+1}) = 0$ or $d(L_{i+1}, L_i) = 0$. We leave it to the reader to verify that any connected generalized metric space corresponds to a connected partially ordered set.

A finite partially ordered set (and the corresponding metric space) can be represented geometrically in a very simple manner. We depict the elements of the partially ordered set as points in three-dimensional space denoted by the same letters as the corresponding elements. We join each pair of points M and N for which $M \bigcirc N$ by a line segment directed from N to M, indicating the direction by an arrow. The geometric figure obtained, consisting of the points (vertices) and the directed line segments joining them, is called a *graph*. We have already seen examples of graphs in figures 9.1 and 9.2.

It is easy to see that if $M \Subset N$, it is possible to travel from N to M by means of a path that moves only in the direction of the arrows.

Metric spaces with nonsymmetric distance functions are especially important in the concept of a discrete topological space.

With this we conclude our study of the concept of distance. We have established that this concept in its many different aspects is connected not only with problems in pure mathematics, but with such practical problems as the construction of error-stabilizing codes. This multiplicity of applications and the complicated logical connections are characteristic of other essential mathematical concepts as well. The principal motivation for the creation of such concepts lies in the possibility of connections and analogies to seemingly unrelated fields and in the need to discover the hidden principles upon which mathematical properties depend.